T0252970

Synthesis Lectures on Engineering, Science, and Technology

The focus of this series is general topics, and applications about, and for, engineers and scientists on a wide array of applications, methods and advances. Most titles cover subjects such as professional development, education, and study skills, as well as basic introductory undergraduate material and other topics appropriate for a broader and less technical audience.

Winfried Mönch

Electronic Structure
of Semiconductor Interfaces

 Springer

Winfried Mönch
Faculty of Physics
Universität Duisburg-Essen
Duisburg, Germany

ISSN 2690-0300 ISSN 2690-0327 (electronic)
Synthesis Lectures on Engineering, Science, and Technology
ISBN 978-3-031-59063-4 ISBN 978-3-031-59064-1 (eBook)
https://doi.org/10.1007/978-3-031-59064-1

© The Editor(s) (if applicable) and The Author(s), under exclusive license to Springer
Nature Switzerland AG 2024

This work is subject to copyright. All rights are solely and exclusively licensed by the Publisher, whether the whole or part of the material is concerned, specifically the rights of translation, reprinting, reuse of illustrations, recitation, broadcasting, reproduction on microfilms or in any other physical way, and transmission or information storage and retrieval, electronic adaptation, computer software, or by similar or dissimilar methodology now known or hereafter developed.
The use of general descriptive names, registered names, trademarks, service marks, etc. in this publication does not imply, even in the absence of a specific statement, that such names are exempt from the relevant protective laws and regulations and therefore free for general use.
The publisher, the authors and the editors are safe to assume that the advice and information in this book are believed to be true and accurate at the date of publication. Neither the publisher nor the authors or the editors give a warranty, expressed or implied, with respect to the material contained herein or for any errors or omissions that may have been made. The publisher remains neutral with regard to jurisdictional claims in published maps and institutional affiliations.

This Springer imprint is published by the registered company Springer Nature Switzerland AG
The registered company address is: Gewerbestrasse 11, 6330 Cham, Switzerland

Paper in this product is recyclable.

To Gisela
with many thanks
for your patience and encouragement

Preface

Interfaces between semiconductors and metals, insulators, or other semiconductors play an important role in almost all semiconductor devices. And, as Kroemer stated quite simply in his *Nobel* Prize Lecture [2001], *"often, it may be said that the interface is the device"*. The research and, shortly after that, the industrial application of semiconductor interfaces started exactly 150 years ago when Braun [1874] published his observation of the *unipolar* conduction of metal contacts on chalcopyrite crystals, as their rectifying behavior was called then in contrast to the ohmic behavior of metals.

First of all, I would like to thank Charles Glaser and Michael Luby, the editors of the book series *Synthesis Lectures on Engineering, Science, and Technology*, for inviting me to write this volume. It is neither another standard review article nor a textbook on the electronic properties of semiconductor interfaces such as volumes 26 and 43 of the *Springer Series in Surface Sciences*, but it is rather designed as a *Lecture*. This volume therefore has a certain personal touch both in the selection of content and in the presentation. I will discuss and analyze some topics that I think still deserve special attention. For example, in my opinion, the best way to understand the electronic properties of metal–semiconductor interfaces is to follow in detail the changes in the electronic properties of clean semiconductor surfaces caused by metal adatoms deposited up to a monolayer coverage and then up to the formation of a thick metal overlayer using many very different measurement methods. The *Synthesis* aspect of the volume results from a comprehensive compilation of experimental barrier heights of metal–semiconductor or *Schottky* contacts and valence-band discontinuities of semiconductor heterostructures and their joint analysis. The conclusion is that the theoretical concept of interface-induced gap states, where the interface dipoles are described by the electronegativity differences of the two solids in contact, consistently and quantitatively explains both the barrier heights of *Schottky* contacts and the band offsets of heterostructures.

Mülheim, Germany
March 2024

Winfried Mönch

Contents

Introduction

1

1.1 Metal–Semiconductor Rectifiers

In his retrospective "*From Solid State Research to Semiconductor Electronics*", Welker (1979) noted that "Solids with semiconductor properties were already playing an important role in scientific research at the beginning of the nineteenth century. This was long before the term "*semiconductor*" was introduced in 1911 in a paper by Koenigsberger and Weiss".

The article "*Über die Stromleitung durch Schwefelmetalle[1]*" published by Braun in 1874 is certainly one of the most important works in this field of research at that time. He wrote[2]: "*Bei einer großen Anzahl natürlicher und künstlicher Schwefelmetalle und sehr verschiedenen Stücken, sowohl Krystallen von so vollkommener Ausbildung, wie ich überhaupt bekommen konnte, als derben Stücken habe ich gefunden, daß der Widerstand derselben verschieden war mit Richtung, Intensität und Dauer des Stromes. Die Unterschiede betragen bis zu 30pCt. des ganzen Werthes. ... Ich habe benutzt Quecksilbercontact, stark gegen gepreßte Kupfer-, Platin- und Silberdrähte und endlich bei einem Stück eine bereits vorhandene Fassung mit dicken Neusilberbügeln.*" Braun substantiated his observations with two data sets, plotted in Fig. 1.1. In contrast to the familiar ohmic behavior of metals, the chalcopyrite crystal clearly shows *unipolar conduction*, as rectification was called at that time (Schuster 1874). This discovery of Braun's caused great interest. Immediately thereafter, Schuster (1874) confirmed it with copper/cuprous oxide point-contacts and Siemens

---

[1] *On the current transport in metal sulfides.*

[2] "In a large number of natural and synthetic sulfides and with very different samples, crystals so perfectly shaped as I could get them as well as rough pieces, I have found that their resistance varied by up to 30% depending on direction, intensity, and duration of the current. ... I have used mercury contacts, wires of copper, platinum, and silver, which were heavily pressed against the samples, and an already existing mounting made of German silver with another piece. ..."

© The Author(s), under exclusive license to Springer Nature Switzerland AG 2024
W. Mönch, *Electronic Structure of Semiconductor Interfaces*, Synthesis Lectures on Engineering, Science, and Technology, https://doi.org/10.1007/978-3-031-59064-1_1

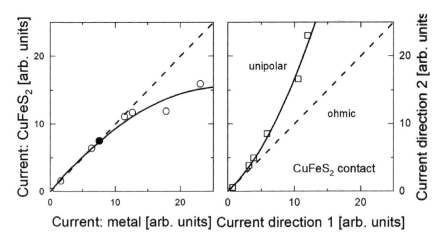

Fig. 1.1 Current flow through a chalcopyrite crystal and a metal resistor and through the forward-and reverse-biased chalcopyrite crystal. The maximum applied voltage was 1 Bunsen (1.8 to 1.9 V). Data from Braun (1874)

(1875) and Adams and Day (1876) with metal-selenium contacts. A few years later, Pickard (1906) and Dunwoody (1906) reported about rectifiers made of silicon and silicon carbide, respectively. Braun (1901 as cited in 1906), Bose (1901/1904), and Pickard (1904/1906) applied metal–semiconductor point-contacts or "cat's whisker" rectifiers as receivers of electro-magnetic radiation.

Planar selenium plate-rectifiers had already been developed by Fritts (1883). However, their commercial implementation was delayed by the development of electron tubes. Fleming (1890) had discovered *"that if a carbon filament electric glow lamp contains a pair of carbon filaments or a single filament and a metallic plate sealed into the bulb, the vacuous space between possesses a unilateral conductivity of a particular kind when the carbon filament, or one of the two filaments, is made incandescent"*. Fifteen years later, Fleming (1905a) then described the properties of respective devices and filed a patent that was granted in the same year (Fleming 1905b). Around 1925, Presser (1925, 1926) and Grondahl (1926) developed Cd/Se and Cu/Cu$_2$O rectifiers, respectively, and forced their industrial production. Even after the invention of silicon rectifiers in the early 1950-s, selenium rectifiers in the form of high-voltage stacks were still produced and used up to the late 1970-s.

Although rectifiers of cuprous oxide and selenium were produced commercially, the production methods were purely empirical. The progress in the physical understanding of metal–semiconductor contacts was very slow. A summary of the many early unsuccessful attempts can be found in Henisch's (1949) book *"Rectifying Semi-Conductor Contacts"*. A huge step forward was eventually made by Schottky and Deutschmann (1929). Their investigations were motivated by the observation of a narrow *"Sperrschicht"* or blocking layer directly at the interface between the mother copper and the Cu_2O overlayer grown on top of it. It was detected by potential measurements using an electrostatic scanning probe while the rectifier was biased in blocking direction.[3] Schottky and Deutschmann (1929) then measured the differential capacitance C of cupreous oxide rectifies as a function of the applied voltage V_a and found the capacitance to decrease with reverse but to increase with forward bias. They modeled the blocking layer by a parallel-plate capacitor, i.e.,

$$C(V_a) \; = \; \varepsilon_b \varepsilon_0 \, A/d(V_a), \tag{1.1}$$

where A and d are the area and the separation, respectively, of the two *virtual* capacitor plates and ε_b is the static dielectric constant of the semiconductor, here Cu_2O. Since the dielectric constant of Cu_2O had not been determined at that time, they assumed $\varepsilon_b \approx 10$, in good approximation to $\varepsilon_b = 7.5$, the value experimentally determined by Kuzel and Weidmann (1970). From the capacitance measured, Schottky and Deutschmann obtained a zero-bias width $d(0) = 3 \times 10^{-5}$ cm of the blocking layer. They further concluded that as a function of the allied voltage V_a the width of the blocking layer increases in reverse but decreases in forward current direction. They then also definitely ruled out current transport by quantum–mechanical tunneling, as the width of the blocking layer is too large for this mechanism to be effective. Schottky and Deutschmann also discussed the possible existence of a space charge in the blocking layer. However, they could not arrive at a consistent model for several reasons.

First, Wilson (1931) did not present his article on *"The Theory of Electronic Semiconductors"* until two years later. Second, the two most important semiconductors at that time, cupreous oxide and selenium were only insufficiently characterized. Already in late 1931, shortly after Wilson's fundamental articles, Schottky and Peierls corresponded on depletion layers, i.e., on the space-charge concept at metal–semiconductor interfaces, as Schottky (1939) mentioned in one of his later articles. However, their conclusions remained inconsistent with experimental results since the sign of the *Hall* coefficient of cupreous oxide had been erroneously given as negative. Fritsch (1935) finally determined the correct, positive sign and thus established Cu_2O as a *p*-type semiconductor. The *p*-type character of selenium was later inferred from the decrease of the barrier heights of metal-selenium rectifiers as a function of the metal work-functions (Schweikert 1939, Schottky 1940). Third, it was important to understand the difference between *chemical* and *physical* blocking layers. At Cu/Cu_2O contacts, the most important type of crystal rectifiers at that time, the oxygen concentration is lower directly at the interface between the copper and the thermally grown cupreous oxide layer. This chemical depletion layer must be considered in addition to the physical depletion or band bending caused by the Cu/Cu_2O barrier height. Schottky (1942) determined the spatial distribution of the acceptors at Cu/Cu_2O interfaces from the above-mentioned capacitance of such rectifiers measured as

---

[3] The measurements are first reported by Schottky and Deutschmann (1929), later published by Schottky et al. (1931) and confirmed by Grondahl (1933).

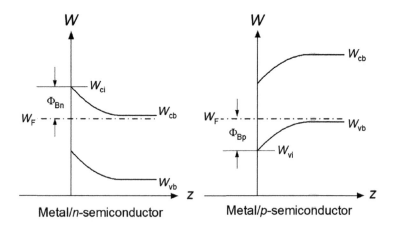

Fig. 1.2 Schematic band diagram of intimate, abrupt, and rectifying metal contacts on n- and p-type semiconductors

a function of the blocking voltage (Schottky and Deutschmann 1929). With selenium rectifiers, on the other hand, only physical depletion layers were considered important.

Finally, Schottky published his *"Halbleitertheorie der Sperrschicht"*[4] in late 1938. He explained the rectification of metal–semiconductor contacts by a band bending in the semiconductor directly at the interface. The band diagrams for n- and p-doped semiconductors are shown schematically in Fig. 1.2. Schottky assumed that the metal–semiconductor work function or, as it is called now, the barrier height, $\Phi_{Bn} = W_{ci} - W_F$ and $\Phi_{Bp} = W_F - W_{vi}$ for n- and p-type semiconductors, respectively, is so large that in thermal equilibrium the blocking layer is depleted of the corresponding mobile majority carriers. Consequently, the respective charged static dopants make up a space charge Q_{sc} with positive and negative sign, respectively, which is balanced by an equivalent charge density Q_m with opposite sign on the metal side, i.e.,

$$Q_m + Q_{sc} = 0. \tag{1.2}$$

An externally applied voltage will either reduce or increase the band bending. Correspondingly, the width of the depletion layer will be decreased and enlarged, respectively. This is schematically explained in Fig. 1.3.

To honor Schottky for his many significant contributions to the fundamental understanding of the electronic properties of metal–semiconductor contacts, they are commonly referred to as *Schottky* contacts.

---

[4] *Semiconductor Theory of the Blocking Layer.*

Fig. 1.3 Band bending at the interface of a metal/ *n*-semiconductor contact in thermal equilibrium ($V_a = 0$) and with forward ($V_a > 0$) and reverse bias ($V_a < 0$) applied (schematically)

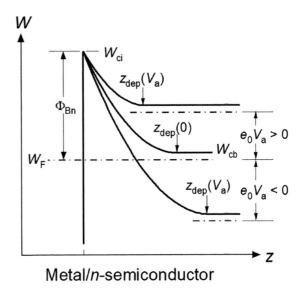

Metal/*n*-semiconductor

1.2 Metal-Induced Gap States

The road to a physical understanding of the band-structure alignment at metal- semiconductor interfaces has been long and thorny. The main steps will be discussed in detail in Sect. 1.3. Only a brief overview will be given here. Mott (1938) was the first to present a simple rule for calculating the barrier heights of *Schottky* contacts. He proposed that the barrier height of *n*-type *Schottky* contacts equals the difference between the work function Φ_m of the metal and the electron affinity χ_s of the semiconductor:

$$\Phi_{Bn} = \Phi_m - \chi_s \qquad (1.3)$$

Schweikert (1939) prepared selenium *Schottky* contacts with 19 different metals and measured their *I/V* characteristics. Schottky (1940) immediately concluded from Schweikert's data that the simple relationship (3), which has entered the literature as the *Schottky–Mott* rule, could not correctly describe the experimental data. The slope parameter $S_\Phi = \partial \Phi_{Bp}/\partial \Phi_m$ was considerably smaller than the value 1 predicted by the *Schottky–Mott* rule. In retrospect this appears quite plausible, since already Mott (1938) had explicitly stated that his result assumed that "*the two* (i.e., the metal and the semiconductor) *do not influence each other in any way*".

Almost a decade later, Bardeen (1947) then proposed that interface states in addition to the space-charge on the semiconductor side of *Schottky* contacts could solve the S_Φ problem. They are an additional charge sink or source that must be accounted for in Eq. 1.1, which then reads.

$$Q_m + Q_s = Q_m + Q_{sc} + Q_{is} = 0 \qquad (1.4)$$

Another two decades later, Cowley and Sze (1965) discussed *Schottky* contacts, which have a continuum of interface states with constant density of states D_{is} across the band gap and a charge neutrality level such that they can accommodate both negative and positive charges, depending on its position relative to the *Fermi* level. They derived the slope parameter

$$S_\Phi = 1/(1 + e_0 \delta_{is} D_{is} / \varepsilon_0 \varepsilon_i), \qquad (1.5)$$

where δ_{is} and ε_i are the effective width of an interfacial layer and the dielectric constant of this layer, respectively. Cowley and Sze made no assumption about the specific character and the origin of these interface states.

Exactly at the same time, Heine (1965) published his fundamental article *"Theory of Surface States"*. Due to the quantum–mechanical tunnel effect, the wave functions of conduction and valence electrons at metal and semiconductor surfaces decay exponentially into vacuum. The behavior at surfaces and interfaces is illustrated in Fig. 1.4. Surface states, and this is also true for adatom-induced surface states, decay exponentially into both vacuum and toward the bulk, but within the solid their charge distribution oscillates. With respect to *Schottky* contacts, Heine concluded that *"for energies in the semiconductor band gap the volume states of the metal all have tails in the semiconductor"*. These metal-induced gap states (MIGS), as they were later called (Louie and Cohen 1976), derive from the complex band structure of the semiconductor and are thus an *intrinsic* property of the semiconductor. They represent the fundamental mechanism that determines the barrier heights of *Schottky* contacts. Heine obtained the barrier height of *n*-type *Schottky* contacts

$$\Phi_{Bn} = \Phi^n_{CNL} + S_\Phi(\varphi_m - \varphi_s) \qquad (1.6)$$

as the sum of the charge-neutrality barrier height $\Phi^n_{CNL} = W_{ci} - W_{CNL}$, where W_{CNL} is the charge-neutrality level of the MIGS continuum, and the dipole $S_\Phi(\varphi_m - \varphi_s)$, where φ_m and φ_s are *"not quite the real work functions, but the volume contributions to them without the surface dipoles on the free surface"*.

At almost the same time, Mead (1966) agued in the same way as Heine with regard to the interface dipole and wrote: *"The work function may be thought of as composed of two parts. One is related to the chemical electronegativity of the metal and the other is a surface dipole layer caused by a distortion of the electronic cloud at the surface. The surface dipole layer is certainly not present in anything like the same form when the metal is in intimate contact with a semiconductor as it is when the metal faces a vacuum. Hence, we will use the electronegativity of the metal as its reference property rather than the work function"*. He thus interpreted Heine's Eq. (1.6) as

Fig. 1.4 Wave-functions of "clean surface" surface states (**a**), of adatom-induced surface states (**b**), and of metal- or, more generally, interface-induced gap states (**c**) (schematic)

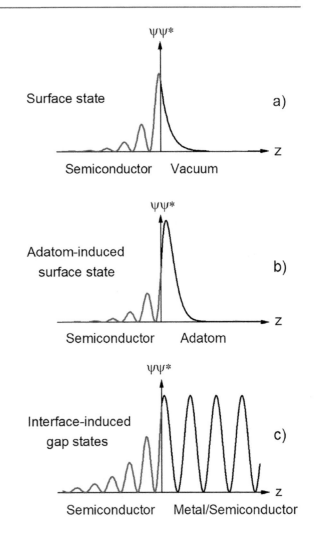

$$\Phi_{Bn} = \Phi^n_{CNL} + S_X(X_m - X_s) \tag{1.7}$$

where X_m and X_s are the electronegativities of the metal and the semiconductor, respectively. However, neither Heine's nor Mead's arguments were generally accepted since a great many physicists, both theorists and experimenters, disliked the concept of electronegativities, and still do, and continue to use the real work functions of metals and the real electron affinities of semiconductors. However, there is a physically seen most simple approach to justify the use of electronegativities in this context.

Adatoms form covalent bonds with surface atoms of semiconductors. The partial ionic character of these bonds causes core-level shifts of the surface atoms which can be measured using X-Ray photoemission spectroscopy (XPS). Mönch (1983) found the experimental adatom-induced shifts reported for C(1s), Si(2p), and Ge(3d) core-levels to depend linearly on the difference of the electronegativities of adatoms and substrate atoms irrespective of the semiconductor considered and whether metal or non-metal adatoms were used. This finding provides a strong experimental argument for using electronegativities instead of real work functions of metals and real electron affinities of semiconductors in expressing the interface dipoles at semiconductor interfaces. This is discussed in detail in Sect. 1.3.

The theorists (Louis and Cohen 1976, Tejedor et al. 1977) immediately accepted Heine's MIGS concept, while the experimenters were very slow to follow. One of the reasons was that Heine's above constraint in relation (1.6) was not considered so that the MIGS concept predicts a linear dependence of the *Schottky* barrier heights on the real work function of the metal. The experimental data available at that time formed rather elongated clouds that could be approximated by linear least-squares fits. However, it took some time to understand that this scatter in the experimental data was caused by lateral inhomogeneities in the barrier heights, which can be easily accounted for, as discussed in Sect. 1.2.

For quite many years, the *Unified Defect Model* of Spicer et al. (1979) then dominated the discussions. While Spicer and coworkers assumed discrete levels of specific native defect, Hasegawa and Ohno (1986) and Walukiewicz (1988, 2001) proposed continua of defect-induced gap levels (DIGS) to determine the barrier heights of *Schottky* contacts. Walukiewicz (1988) recognized "*a striking correlation between the Fermi level in heavily radiation damaged semiconductors and at metal–semiconductor interfaces*", and he claimed that "*the correlation provides critical evidence supporting the defect model for Schottky-barrier formation*". However, there is no experimental evidence that native defects determine the barrier heights of *Schottky* contacts, but contrary evidence. Grazing incidence X-ray diffraction demonstrated that under thick Ag and Pb films evaporated on Si(111)-7 × 7 surfaces the 7 × 7 reconstruction persists [Hong et al. 1992, Howes et al. 1995]. Medium-energy ion channeling experiments gave upper limits of 1×10^{13} and 3×10^{13} Si atoms per cm^2, i.e., less than 1 and 4% of the atoms in a bulk Si{111} layer being displaced from lattice sites in type-A and type-B epitaxial NiSi$_2$/Si(111) contacts, respectively (Vrijmoeth et al. 1990).

It is worth mentioning already here that Brudnyi et al. (1995) also noticed the correlation between the "*stabilization of the Fermi level in highly irradiated semiconductors*" and the position of the charge-neutrality level at semiconductor interfaces. However, Brudnyi et al. identified this as a *mere correlation* and correctly attributed it to the common origin of radiation-induced gap states of native defects in the bulk and of interface-induced gap states: they both originate from the complex band structure of the semiconductors. This is discussed in detail in Chap. 6.

Tersoff (1984a, 1986) was then the first to calculate the charge-neutrality levels or, as they were later called, branch-point energies of the virtual gap states of the complex band structures of the elemental semiconductors Si and Ge, and a total of 12 of the III-V and

II-VI compounds which are entered in Table 4.1. He compared his calculated *"canonical barrier heights"* with experimental barrier heights of corresponding Au *Schottky* contacts and found good agreement. With respect to the analysis of experimental data in Sect. 1.2, it is worth mentioning that he considered *C/V* barrier heights only. This comparison is reasonable because the corresponding dipole terms $S_X(X_{Au}-X_s)$ only vary between 0.02 and 0.08 eV. Unfortunately, he did not discuss the experimentally observed *"deviations from the canonical barrier heights calculated"* but acknowledged that they *"may be attributed, at least in part, to the M-S electronegativity difference"*.

In a lecture commemorating Schottky's 100th birthday, Mönch (1986) re-analyzed the experimental slope parameters S_Φ collected by Kurtin et al. (1969) and revised by Schlüter (1978). It is worth mentioning, that Schlüter not only revised the S_Φ data but also converted them to S_X values using Pauling's electronegativities. Considering Maue's (1935) early treatment of the complex band structure of 1-dimesional solids in relation (1.5), it is seen that both the density of states D_{is} and the effective width δ_{is} of the interfacial layer depend on the width of the band gap, which Mönch equated with Penn's (1962) average or dielectric band gap

$$ W_{dg} \, = \, \hbar\omega_p/(\varepsilon_\infty - 1)^{1/2}, \tag{1.8} $$

where $\hbar\omega_p$ is the plasmon energy of the bulk valence electron. The result was that the slope parameter depends directly on the optical dielectric susceptibility $\varepsilon_\infty-1$ of the respective semiconductor. In Fig. 1.5, the slope parameters reported by Schlüter (1978), Pong and Paudyal (1981), and Jacob et al. (1987) are plotted versus $\varepsilon_\infty - 1$. The linear least-squares fit

$$ A_X/S_X - 1 \, = \, 0.1(\varepsilon_\infty - 1)^2, \tag{1.9} $$

where $A_X = 1.79$ eV/Pauling unit, describes the slope parameters of all *Schottky* contacts irrespective of whether they are *traditional* semiconductors, insulators such as Al_2O_3 or SiO_2, ionic crystals such as LiF and BaF_2, or *van der Waals*-bonded solid Xenon. The common feature of these solids is the energy gap between valence- and conduction band.

Inspired by Tersoff's results, Mönch (1996) calculated an extended set of branch-point energies including diamond, and in particular, silicon carbide and the group-III nitrides. He applied Baldereschi's (1973) concept of mean-value $\underline{k}$-points in the *Brillouin* zone and again Penn's (1962) concept of average or dielectric band gaps together with Maue's (1935) early treatment of the complex band structure of 1-dimesional solids and used the empirical tight-binding approximation to calculate the valence-band dispersion at the mean-value $\underline{k}$-point. In this way, he avoided using the *scissors operator* to adjust the width of the calculated band gap to the real value, as Tersoff had done. His data can also be

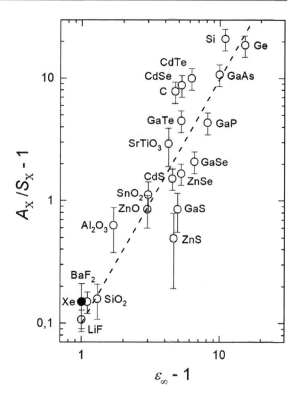

Fig. 1.5 Slope parameters S_X $= d\Phi_{Bn}/dX_m$ as a function of the optical susceptibility $\varepsilon_\infty -1$ of the semiconductors. The dashed line is the linear least-squares fit to the data. From Mönch (1986)

found in Table 4.1. As shown in detail in Sect. 1.3, Mönch's branch-point energies have been excellently confirmed by the computationally demanding quasiparticle GW band-structure calculations of *Bechstedt* and co-workers (Schleife et al. 2009, Höffling et al. 2010, Belabbes et al. 2011, 2012) and by the recent elaborate work of Guo et al. (2019), Ghosh et al. (2022), and Varley et al. (2024).

1.3   Semiconductor Heterostructures

Commercial selenium rectifiers were made by depositing Cd on thick Se films. Poganski (1952, 1953) observed that during the processing of such rectifiers Cd and Se react so that no metal–semiconductor contacts are formed, but *n*-CdSe/*p*-Se heterostructures. The importance of heterostructures in today's semiconductor devices became widely known when Kroemer and Alferov were awarded the *Nobel* Prize in Physics in 2000. Kroemer (1957) had presented the *Theory of Wide-Gap Emitters for Transistors* and both Kroemer (1963) and Alferov and Kazarinov (1963) had independently proposed the double-heterostructure concept of injection lasers. In their *Nobel* Prize lectures, Alferov (2001) and Kroemer (2001) stated that "*It is impossible to imagine now modern solid-state*

Fig. 1.6 Schematic energy-band diagram of semiconductor heterostructures. The band bending due to the doping of the semiconductors is not considered

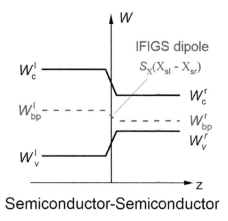

Semiconductor-Semiconductor

physics without semiconductor heterostructures" and that *"often, it may be said that the interface is the device"*, respectively.

Anderson (1962) was the first to propose an explanation for the band alignment in heterostructures. Following the *Schottky–Mott* model for *Schottky* contacts, he equalized the vacuum levels of two semiconductors without considering any surface or interface states. Thus, he obtained the valence-band offset $\Delta W_v = W_{vil} - W_{vir}$ as the difference of the ionization energies $I_{l,r} = W_{vac} - W_{vil.r}$ of the two semiconductors, where l and r stand for the left and the right side of the heterostructure, respectively. Tejedor and Flores (1978) then applied Heine's (1965) concept of interface-induced gap states also to heterostructures. They obtained the valence-band offset as

$$\Delta W_v = \Phi^p_{CNLl} - \Phi^p_{CNLr} + S_\Phi(I_l - I_r), \tag{1.10}$$

i.e., as the difference of the IFIGS CNL-s of the two semiconductors plus an electric dipole term $S_\Phi(I_l - I_r)$. Unfortunately, Tejedor and Flores overlooked Heine's (1965) argument that at interfaces the volume contributions, here of the ionization energies, rather than their real values should be considered, and the same also applies to Mead's (1968) suggestion to use electronegativities instead. This means that by analogy to *Schottky* contacts, where Eq. (1.6) was replaced by Eq. (1.7),

$$\Delta W_v = \Phi^p_{bpl} - \Phi^p_{bpr} + S_X(X_l - X_r). \tag{1.11}$$

should be used here instead of Eq. (1.10). This is explained schematically in Fig. 1.6.

Experimental Data Base

2.1 I/V Characteristics of *Schottky* Contacts: Barrier Height and Ideality Factor

The current transport in metal–semiconductor interfaces was first analyzed in detail by Bethe (1942). He formulated the conditions under which the current flow occurs via thermionic emission over the barrier. His criteria are generally satisfied by the metal–semiconductor contacts considered in this lecture. Tunneling through the top of the *Schottky* barrier was already ruled out by Schottky and Deutschmann (1929). For moderate doping levels, the current/voltage characteristics (*I/V*) of *n*-type *Schottky* contacts can be written as

$$J_{te} = AA_R^* T^2 \exp(-\Phi_{Bn}/k_B T)[\exp(e_0 V_c/k_B T)) - 1], \tag{2.1}$$

with the effective *Richardson* constant

$$A_R^* = A_R \frac{m_n^*}{m_0}, \tag{2.2}$$

where $A_R = 4\pi e_0 k^2_B m_0/h^2 = 120\text{A cm}^{-2}\text{K}^{-2}$ is the *Richardson* constant for thermionic emission of nearly free electrons into vacuum (Richardson 1914) and m_n^* is the effective mass of the electrons in the conduction band of the semiconductor. For derivation of relation (2.1), the reader is referred to the books by Rhoderick and Williams (1988) or Mönch (2004).

As with the thermionic emission of electrons from metals (Schottky, 1914), the image- force concept must also be considered at metal–semiconductor interfaces. A metal strongly distorts the radially symmetric *Coulomb* field of a point charge in front of it. The surface of the metal is an equipotential surface because no current flows parallel to it,

© The Author(s), under exclusive license to Springer Nature Switzerland AG 2024
W. Mönch, *Electronic Structure of Semiconductor Interfaces*, Synthesis Lectures
on Engineering, Science, and Technology, https://doi.org/10.1007/978-3-031-59064-1_2

Fig. 2.1 Band bending at the interface of a metal and an n-type semiconductor in thermal equilibrium ($V_c = 0$) and with forward bias ($V_c > 0$) applied (schematically)

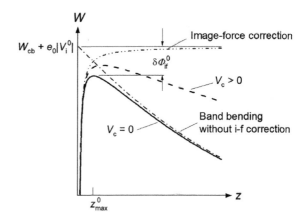

so the electric field must be perpendicular to it. This requirement is easily satisfied by a mirror image of the original point charge, which has the same size but the opposite sign.

The attractive interaction between a real point charge and its image induced in the metal lowers the potential energy of the electron. For vacuum tubes, Schottky (1914) deduced a lowering of the metal work-function by the image-force effect. At metal–semiconductor interfaces, the image-force effect then analogously decreases the barrier height. This is explained in Fig. 2.1. For an n-type semiconductor, the image-force lowering of the barrier height results as (see, for example, Mönch (2004)).

$$\delta\Phi_{if} = e_0\{(2e_0^2 N_d/(4\pi)^2\varepsilon_\infty^2\varepsilon_b\varepsilon_0^3)[e_0(|V_i^0 - V_c| - k_B T\}^{\frac{1}{4}}, \qquad (2.3)$$

where N_d is the donor density, and ε_b and ε_∞ are the static and the optical dielectric constants, respectively, of the semiconductor.

Because of the image-force effect, the effective barrier heights even of ideal metal–semiconductor contacts depend on the externally applied voltage. Quite generally, the bias dependence of the barrier height may be expressed as

$$\Phi_{Bn} = \Phi_{Bn}^0 + \beta e_0 V_c + \cdots = \Phi_{Bn}^0 + \left(1 - \frac{1}{n}\right) e_0 V_c + \ldots, \qquad (2.4)$$

where

$$\Phi_{Bn}^0 = \Phi_{Bn}^{hom} - \delta\Phi_{if}^0 \qquad (2.5)$$

is the zero-bias barrier height and Φ_{Bn}^{hom} is the barrier height of the *Schottky* contact if the zero-bias image-force lowering $\delta\Phi_{if}^0$ is not considered. As in relation (2.4), the zero-bias derivative $\beta = \partial\Phi_{Bn}/\partial e_0 V_c$ can be replaced by an ideality factor n. Because of the image-charge effect, the ideality factor

$$n_{if} = \left(1 - \delta\Phi_{if}^0/4e_0\left|V_i^0\right|\right)^{-1} \tag{2.6}$$

describes the voltage-dependence of barrier heights of *ideal Schottky* contacts. The combination of relations (2.1), (2.4), and (2.5) finally yields the I/V characteristic of *ideal Schottky* contacts

$$J_{te} = A_R^*T^2\exp\left(-\Phi_{Bn}^0/k_BT\right)\exp(e_0V_c/n_{if}k_BT)[1 - \exp(-e_0V_c/k_BT))]. \tag{2.7}$$

In *real Schottky* diodes, other effects than the image force reduction can additionally cause a bias dependence of the barrier height. The most important case are patches of reduced barrier height with lateral dimensions smaller than the depletion layer width that will be discussed below in Sect. 2.4. Consequently, n_{if} will be replaced by a more general ideality factor n so that the I/V characteristic (2.7) becomes

$$J_{te} = A_R^*T^2\exp\left(-\Phi_{Bn}^0/k_BT\right)\exp(e_0V_c/nk_BT)[1 - \exp(-e_0V_c/k_BT))]$$

$$= J_0\exp(e_0V_c/nk_BT)[1 - \exp(-e_0V_c/k_BT))] \tag{2.8}$$

with the saturation current density

$$J_0 = A_R^*T^2\exp(-\Phi_{Bn}^0/k_BT) \tag{2.9}$$

It was found early on that relation (2.1) did not accurately describe the forward current–voltage characteristics of *real Schottky* contacts. Therefore, the slope parameter of relation (2.1) was *adjusted* by an ideality factor defined as

$$1/n = (k_BT/e_0)(d(InJ)/dV_c) \tag{2.10}$$

and which was then simply and formally added to the exponential in the rectification bracket as follows

$$J_te = A_R^*T^2exp(-\Phi_{Bn}(k_BT)[exp(e_0V_c)/(nk_BT)) - 1]. \tag{2.11}$$

No *physical* explanation for this procedure was ever given, i.e., the ideality factor was used as an *unexplained fitting* parameter rather than a measure of the voltage dependence of the barrier height. Already in 1988, Rhoderick and Williams explicitly pointed out that *"this form is incorrect because the barrier lowering affects the flow of electrons from metal to semiconductor as well as the flow from semiconductor to metal, so that the second term in the square bracket must contain n".* Nevertheless, relation (2.11) is still used in most articles reporting on and analyzing *I/V* characteristics of *Schottky* contacts. This is the reason for the preceding, somewhat lengthy discussion of the ideality factor.

2.2   C/V Characteristics of *Schottky* Contacts: Flat-Band Barrier Height

The space charge and the width of the depletion layer in metal–semiconductor contacts both vary as a function of the externally allied voltage. Figure 2.2 explains this behavior. Schottky (1942) derived the differential capacitance per unit area of depletion layers in metal–semiconductor contacts as

$$C_{dep} = \partial Q_{sc}/\partial V_c = \{e_0^2 \varepsilon_b \varepsilon_0 N_d / [2e_0(|V_i^0| - V_c) - k_B T]\}^{\frac{1}{2}}. \qquad (2.12)$$

The additional term $k_B T$ accounts for the penetration of electrons from the bulk into the depletion layer. The inverse square of the $1/C_{dep}^2$ of the depletion layer capacitance per unit area thus varies proportional to the applied voltage, and the slope

$$d(1/C_{dep}^2)/dV_c = -2/(e_0 \varepsilon_b \varepsilon_0 N_d) \qquad (2.13)$$

Fig. 2.2 Band bending at the interface of a metal contact and an *n*-type semiconductor in thermal equilibrium ($V_c = 0$) and with forward ($V_c > 0$) and reverse bias ($V_c < 0$) applied (schematically). The image-force effect is not considered

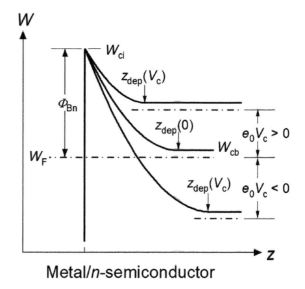

is determined by the distribution $N_d(z)$ of the donors. The extrapolated intercepts V_a^{ex} on the abscissa of plots of $1/C_{dep}^2$ versus V_c are then equal to $(e_0|V_i^0| - k_BT)/e_0$. The extrapolation is equivalent to flat bands up to the interface. Thus, the extrapolated intercept gives the C/V or flat-band barrier height of the *Schottky* contact

$$\Phi_{Bn}^{fb} = e_0|V_i^0| + (W_{cb} - W_F) = e_0 V_a^{ex} + k_BT + (W_{cb} - W_F). \qquad (2.14)$$

For completely ionized donors, the position of the Fermi level below the bulk conduction-band minimum results as

$$W_{cb} - W_F = k_BT \ln(N_c/N_d), \qquad (2.15)$$

where N_c is the effective density of states of the conduction band.

2.3   I/V and C/V Characteristics of Ag/n-Si(111) Schottky Contacts

2.3.1   Three Mechanisms

As an example, experimental data of Ag/n-Si(111) Schottky contacts are first discussed below. These contacts are chosen because silicon surfaces have been extensively studied, and specific techniques have been developed to prepare clean metal-silicon interfaces. For more details the reader is referred to Sect. 11.1 and 10.1 of the monographs by Mönch (2001) and (2004), respectively.

Figures 2.3 and 2.4 display I/V and C/V characteristics of Ag/n-Si(111) *Schottky* contacts with two different interface structures, namely unreconstructed Si(111)–(1 × 1)[i] and reconstructed Si(111)–(7 × 7)[i]. The Si(111)–(7 × 7)[I] interface structure is discussed in Sect. 2.5.

For thermionic-emission over the barrier, relation (2.8) predicts a linear dependence of $\ln[J_{te}/(1-\exp(-e_0V_c/k_BT)]$ as a function of the voltage V_c across the depletion layer (Missous & Rhoderick, 1986). The slope parameter and the intercept should then equal nk_BT and $A_R^*T^2\exp(-\Phi_{Bn}^0/k_BT)$, respectively. Figure 2.5 shows the corresponding plots of the I/V data shown in Fig. 2.3. The dashed lines are linear least-squares fits to the data points. The linear regression coefficients $r_1 = 0.99996$ demonstrate the very good quality of both I/V characteristics. The slight deviations from the straight lines for the largest voltages applied in forward direction are due to the series resistance of the diodes. Hence, the error is small when the applied voltage is assumed to drop completely across the depletion layer. The slope parameters and the intercepts of the linear least-squares fits yield ideality factors of 1.08 ± 0.001 and 1.05 ± 0.001 and *effective* zero-bias barrier heights of 0.74 and 0.69 eV for the diodes with (1 × 1)[i] and (7 × 7)[i] interface structure, respectively. These n values are significantly larger than the image-force value $n_{if} = 1.01$

Fig. 2.3 Current–voltage
characteristics of forward- and
reverse-biased Ag/
n-Si(111)–(7 × 7)^i and –(1 ×
1)^i *Schottky* diodes. The diode
areas are 6.3×10^{-3} and $6.4 \times$
10^{-3} cm^2, respectively. From
Schmitsdorf (1993)

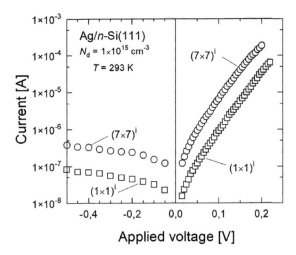

Fig. 2.4 Differential
capacitance of one Ag/
n-Si(111)–(7 × 7)^i and Ag/
n-Si(111)–(1 × 1)^i diode each
as a function of the applied
bias in reverse direction. The
diodes had an area of $7.3 \times$
10^{-3} cm^2. The dashed lines are
linear least-squares fits to the
data points. From Schmitsdorf
et al. (1995)

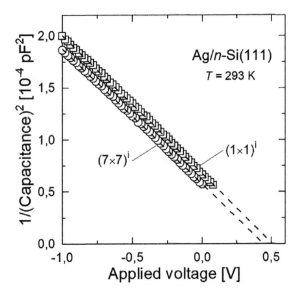

obtained from relation (2.6). Thus, they indicate the presence of extrinsic mechanisms in
addition to the image force that cause the barrier height to depend on the bias.

The linear least-squares fits to the experimental *C/V* data of the Ag/n-Si(111)–(1 ×
1)^i and –(7 × 7)^i diodes (Schmitsdorf 1993) plotted in Fig. 2.4 prove that the donor
densities in the depletion layers of both diodes are constant and identical, $N_d = 1.7 \times$
10^{15} cm^{-3}. At room temperature, the donors are completely ionized, and relation (2.14)
gives the corresponding energy position $W_{cb} - W_F = 0.24$ eV of the Fermi level below
the conduction-band edge in the bulk. With the extrapolated abscissa intercepts $V_a^{ex} =$

Fig. 2.5 Logarithmic plot of *I/*
[1-exp(-e₀Vₐ/k_BT)] as a
function of the applied voltage
V_a for Ag/*n*-Si(111)–(1 × 1)ⁱ
and –(7 × 7)ⁱ *Schottky* diodes.
The data are the same as in
Fig. 2.3. From Mönch (2004)

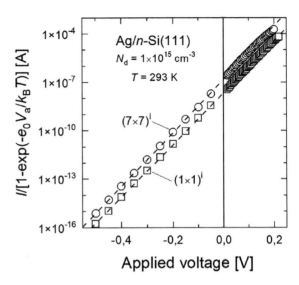

0.51 V and 0.44 V of the (1 × 1)ⁱ and the (7 × 7)ⁱ diode, respectively, relation (2.14) then gives the flat-band barrier heights $\Phi_{Bn}^{fb} = 0.78$ eV and 0.71 eV of the Ag/*n*-Si(111)–(1 × 1)ⁱ and the -(7 × 7)ᴵ *Schottky* contacts, respectively.

The histograms presented in Fig. 2.6 show that the ideality factors and zero-bias barrier heights of *Schottky* contacts fabricated under experimentally identical conditions vary from one diode to the next. Thus, the transport properties of *each single Schottky* contact are characterized by two *phenomenologically* defined parameters: its *effective* (zero-bias) barrier height Φ_{Bn}^{eff} and its ideality factor *n*. Figure 2.7 shows that the *effective* (zero-bias) barrier heights Φ_{Bn}^{eff} and the ideality factors *n* vary in a correlated manner.

The experimental data shown in Fig. 2.7 suggest three distinct mechanisms (Schmitsdorf et al. 1995). The first one is the linear correlation

$$\Phi_{Bn}^{eff}(n) = \Phi_{Bn}^{nif} - \varphi_p(n - n_{if}). \tag{2.16}$$

The resulting characteristic parameters are now the barrier heights $\Phi_{Bn}^{nif}(1 \times 1)^i$ and $\Phi_{Bn}^{nif}(7 \times 7)^i$ of the Ag/*n*-Si(111) contacts with unreconstructed and reconstructed interfaces, respectively, at the ideality factor n_{if} which is only determined by the image-force lowering. Since the image-force effect is always present in real *Schottky* contacts, the ideality factor n_{if} is the appropriate reference rather than $n = 1$, which can, however, be used as a suitable approximation.

The second mechanism is the offset $\delta\Phi_{Bn}^{(7 \times 7)i} = \Phi_{Bn}^{nif}(1 \times 1)^i - \Phi_{Bn}^{nif}(7 \times 7)^i = 47$ meV of the barrier heights of the (7 × 7)ⁱ-⁻reconstructed with respect to the (1 × 1)ⁱ-unreconstructed interface at the ideality factor n_{if}.

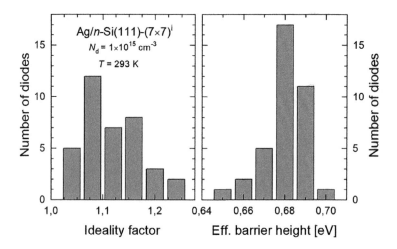

Fig. 2.6 Histograms of ideality factors and effective barrier heights determined from *I/V* characteristics of identically fabricated Ag/n-Si(111)–(7 × 7)[i] diodes at room temperature. Data from Schmitsdorf et al. (1995)

Fig. 2.7 Effective barrier heights versus ideality factors determined from *I/V* curves of Ag/*n-Si*(111)–(7 × 7)[i] and -(1 × 1)[i] contacts at room temperature. The data of the diodes with (7 × 7)[i]-reconstructed interfaces are the same as in Fig. 3.6. The data of Fig. 2.4 are marked in red. The dashed and dash-dotted lines are linear least-squares fits to the data. From Schmitsdorf et al. (1995)

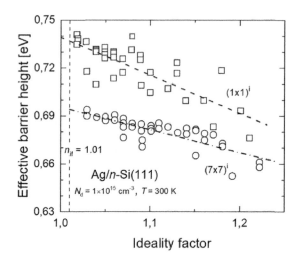

Finally, the third mechanism determines the barrier height of the of the Ag/*n*-Si(111)-$(1 \times 1)^{I}$ contact with the unreconstructed $(1 \times 1)^{I}$ interface. While the first two mechanisms are of extrinsic origin, the third one is likely to be *the* intrinsic one (Mönch, 1999). The three mechanisms are discussed in detail below.

The correlation (2.16) between the effective barrier heights $\Phi_{Bn}^{eff}(n)$ and the ideality factors n of a series of *Schottky* contacts fabricated under identical conditions suggest the following two comments. For further physical discussions of such data sets, it does not seem reasonable, to select a "*best value*" diode with the largest barrier height and the smallest ideality factor or, second, to consider *averages* of the effective barrier heights and ideality factors.

2.4   Laterally Inhomogeneous or "Patchy" *Schottky* Contacts

Ideality factors of *Schottky* contacts larger than n_{if} imply a pronounced dependence of their barrier heights on the applied voltage. Patches with reduced barrier height and lateral dimensions smaller than the width of the depletion layer are the only known model that explains this behavior. Here, the specific mechanism causing the barrier-height lowering needs not be known in detail.

Freeouf et al. (1982) were the first to realize that the potential beneath such patches exhibits a saddle point. This is illustrated in Fig. 2.8. The thermal emission over the saddle points of such patches represents the preferred current paths in "patchy" *Schottky* contacts. Freeouf et al. already found that the saddle point potentials and thus the effective barrier heights of "patchy" *Schottky* contacts vary as a function of the applied voltage (see Fig. 2.9).

Their simulations of current transport through such patches also demonstrated that the effective barrier heights decrease, and the ideality factors increase as they reduced the lateral dimensions of the patch. Figure 2.10 shows some of their results, which they tabulated only, as well as results from the later analytical small-signal approach of Tung (1992). Unfortunately, Freeouf et al. (1982) did not mention that Ohdomari et al. (1978) had already observed that the effective barrier heights of their Ir/n-Si contacts became smaller with increasing ideality factors. Wagner et al. (1983) plotted the effective barrier heights of their PtSi/n-Si contacts as a function of their ideality factors and observed the effective barrier heights to decrease with increasing ideality factors. Wagner et al.

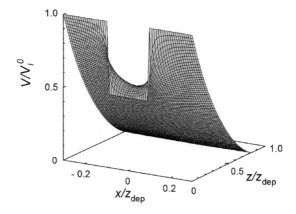

Fig. 2.8 Calculated potential distribution underneath and around a patch of reduced interface potential embedded in a region of larger interface band-bending. The lateral dimension and the interface potential reduction of the patch are set to two tenths of the depletion layer width z_{dep} and one half of the interface potential of the surrounding region. From Mönch (2004)

Fig. 2.9 Potential distribution below the center of a circular patches with a barrier height lowered by $\Delta_\pi = 0.35$ eV and with a radius of $R_\pi = 10$ nm embedded in a region of constant barrier height of $\Phi_{Bn}^{hom} = 0.85$ eV and a depletion layer width of 103 nm as a function of the applied voltage across the depletion layer. From Mönch (2004)

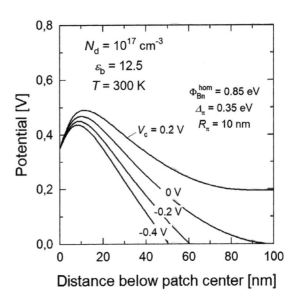

Distance below patch center [nm]

additionally determined the flat-band barrier heights from the *C/V* characteristics of their contacts and found that the barrier height obtained by extrapolating the Φ_{Bn}^{eff}-versus-*n* plot to $n = 1$, what they called the *fundamental* barrier height, agrees with the flat-band barrier height determined from the *C/V* characteristics. Wagner et al. (1983) already concluded and generalized *"that the (extrapolated) fundamental barrier height and not the (effective) zero-voltage barrier heights (of individual diodes) provide a better characterization of the physical barrier at the metal–semiconductor junction"*. However, they did not consider the simulations of patchy *Schottky* contacts by Freeouf et al. (1982). Unfortunately, the findings of *Wagner* et al. did not attract the attention of the scientific community at that time.[1]

The results on patchy *Schottky* contacts by Freeouf et al. (1982) were not appreciated by the scientific community at first since these authors had discussed them exclusively as an explanation for their model of *"mixed phase contacts isolated from one another"*. This model (Freeouf & Woodall, 1981) assumes that patches of anion-rich phases exist at interfaces between metals and both III-V and II-VI compounds. Obviously, such phases are not conceivable for the elementary semiconductors diamond, Si, and Ge. The situation changed completely when Tung (1992) published his extended analytical study of the potential and the electron transport at laterally inhomogeneous, nondegenerate *Schottky* contacts. He confirmed the results obtained by Freeouf et al. (1982) for single patches and extended his calculations to patch densities with *Gaussian* distributions of their characteristic parameters

---

[1] The author first became aware of this article of Wagner et al. [1983] during a literature search in connection with this publication.

Fig. 2.10 Effective barrier height as a function of the ideality factor obtained from simulated current–voltage characteristics of laterally inhomogeneous metal/n-Si contacts. The contacts possess one stripe with a reduced barrier height of 0.4 eV and lateral widths of 0.0313, 0.0625, 0.125, 0.25, 0.5, and 1 μm. The □ data are from Freeouf et al. (1982) and the dashed line is a result of an analytical solution for circular patches (Tung, 1992). From Mönch (2004)

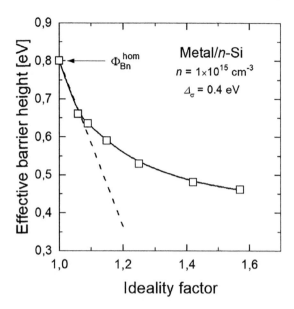

$$\gamma = 3(\Delta_\pi R_\pi / 4)^{(1/3)}, \tag{2.17}$$

where Δ_π and R_π are the barrier-height lowering and the radius, respectively, of circular patches.

Olbrich et al. (1997, 1998) fabricated defined inhomogeneous *Schottky* contacts by first growing Co clusters of different diameters on n-GaAs$_{0.67}$P$_{0.33}$(001) surfaces, which were then embedded in Au films. They determined the potential profiles under and around the Co patches using ballistic-electron-emission-microscopy (BEEM). The basic idea of BEEM is that monoenergetic electrons injected by a metal tip through the vacuum gap into the thin metal film of a *Schottky* contact lose no energy on their way through the metal to the metal–semiconductor interface and are collected as ballistic electrons by the semiconductor.[2] Bell and Kaiser (1988) derived the BEEM current–voltage dependence as

$$I_{coll} = R^* I_{tip} (e_0 V_{tip} - \Phi_{Bn})^2, \tag{2.18}$$

where I_{coll} and I_{tip} are the BEEM collector tip current, respectively, and V_{tip} the voltage applied between the tip and the metal film. BEEM can be used to determine local barrier heights of *Schottky* contacts with a lateral resolution of 1 nm.

Figure 2.11a displays the barrier-height profiles below and around Co clusters of different diameters as obtained by Olbrich et al. (1997) from their experimental I_{coll}/V_{tip}

---

[2] For more details the reader is referred to Sect. 3.6 of the monograph Mönch [2004] and references cited therein.

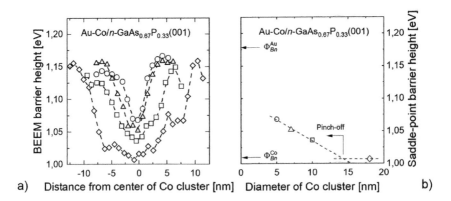

Fig. 2.11 BEEM barrier-height line-scans (**a**) and saddle-point barrier heights (**b**) underneath Co clusters embedded in Au films on an n-GaAs$_{0.67}$P$_{0.33}$ substrate. Diameter of Co clusters: ($\bigcirc$) 5 nm, ($\triangle$) 7 nm, ($\square$) 10 nm, ($\diamond$) 18 nm. Data from Olbrich et al. (1997, 1998)

characteristics. The data plotted in Fig. 2.11b reveal that pinch-off occurs at Co patches with diameters less than approximately 15 nm. Tung (1992) derived the condition for pinch-off at circular patches with diameters

$$2R_\pi < \frac{\Delta z_{dep}}{e_0 V_i^0}. \tag{2.19}$$

Using the respective data of the Au–Co/n-GaAs$_{0.67}$P$_{0.33}$(001) patches, relation (2.19) yields a critical patch diameter of 16 nm for pinch-off. This value is in perfect agreement with the experimental result.

Already prior to Tung's article, Song et al. (1986) and Werner and Güttler (1991) had proposed a somewhat simpler model of the electronic transport across inhomogeneous *Schottky* contacts. They assumed *Gaussian* distributions of barrier-height inhomogeneities, which, and that is most important, occur on a length scale small compared to the width of the space-charge layer. This approach of Song et al. and Werner and Güttler is widely used in the analysis the temperature dependence of *I/V* characteristics of *Schottky* contacts. Their model replaces the "*effective*" zero-bias barrier height Φ_{Bn}^{eff} in relation (2.8) by an "*apparent*" zero-bias barrier height

$$\Phi_{Bn}^{app} = \Phi_{Bn}^{hom}(0) - \frac{e_0 \sigma_s^2}{2k_B T}, \tag{2.20}$$

where σ_s is the standard deviation of the assumed *Gaussian* distribution. The maximum value $\Phi_{Bn}^{hom}(0)$ of the *Gaussian* distribution is interpreted as the barrier height of the laterally homogeneous *Schottky* contact at zero temperature.

Fig. 2.12 Histogram of BEEM barrier heights of a Au/n-Si(001) contact with a donor density 1.2×10^{15} cm^{-3}. The dashed curve is a *Gaussian* least-squares fits to the data. The center and the standard deviation of the *Gaussian* are 0.814 ± 0.0002 eV and 0.008 ± 0.0001 eV, respectively. Data from Morgan et al. (1996)

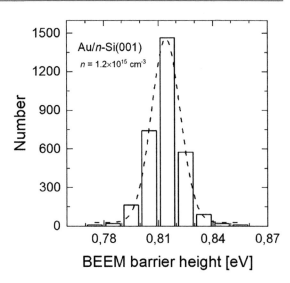

Gaussian distributions of barrier height inhomogeneities have been experimentally seen in many BEEM studies. Morgan et al. (1996), for example, fabricated Au/n-Si(001) *Schottky* contacts and determined the lateral distributions of BEEM barrier heights as well as their *I/V* barrier heights and ideality factors. Figure 2.12 displays their results obtained with an Au/n-Si(001) contact. The center of the *Gaussian* least-squares fit to the BEEM data is 0.814 eV, and the *I/V* barrier height and ideality factor reported are 0.91 eV and 1, respectively.

A standard procedure in the analysis of *I-V-T* data of *Schottky* contacts is the use of *Richardson* plots. Relation (2.9) can be rewritten as

$$\ln(J_0/T^2) = \ln\left(A_R^*\right) - \frac{\Phi_{Bn}^0}{k_B T}. \tag{2.21}$$

The intercept and the slope of a *Richardson* plot $\ln(J_0/T^2)$-versus-$1/T$ give the *Richardson* constant A_R^* and the zero-bias barrier height Φ_{Bn}^0, respectively, of the *Schottky* contact, but only, if the barrier height does not depend on temperature. This basic requirement is not met by *real Schottky* contacts because of lateral inhomogeneities their effective barrier heights vary as a function of temperature. As an example, Fig. 2.13b shows respective data of an Au/n-Si(001) contact. Therefore, such *conventional Richardson* plots give no relevant information for real *Schottky* contacts. If we assume with Song et al. (1986) and Werner and Güttler (1991) a *Gaussian* distribution of barrier-height inhomogeneities, the zero-bias barrier height Φ_{Bn}^0 must be replaced in relation (2.19) by the apparent barrier height (2.18) which then yields

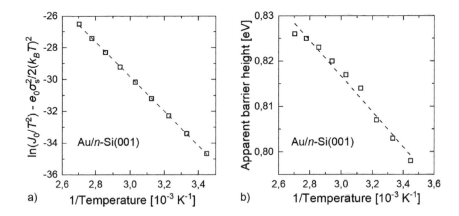

Fig. 2.13 Modified *Richardson* plot (**a**) and apparent barrier heights (**b**) of an Au/n-Si(001) Schottky contact as a function of the reciprocal temperature. Data from Mahato and Puigdollers (2018)

$$\ln(J_0/T^2) = \ln\left(A_R^*\right) - \Phi_{Bn}^{hom}(0)/k_B T + e_0 \sigma_s^2/2(k_B T)^2. \qquad (2.22)$$

In the resulting *"modified" Richardson* plots, $\ln(J_0/T^2) - e_0\sigma_s^2/2(k_B T)^2$ is then plotted versus $1/T$.

As an example, the *I-V-T* data of Au/n-Si(001) *Schottky* contacts as reported by Mahato and Puigdollers (2018) are discussed below. Unfortunately, experimental studies with Ag/p-Si *Schottky* contacts by Acar et al. (2004) and Kumar et al. (2019) cannot be considered here because the barrier heights reported there are very large and completely inconsistent with the data of the Ag/n-Si contacts analyzed above. The arguments for this statement are presented in Sect. 4.1.1. Most probably, the discrepancies are due to details in the fabrication of the Ag/p-Si contacts, which most likely hindered the formation of intimate Ag-Si interfaces by the presence of residual oxide layers.

Mahato and Puigdollers (2018) analyzed their experimental data using the approach of Song et al. (1986) and Werner and Güttler (1991). Figure 2.13a displays a modified *Richardson* plot of their respective data. Considering relation (2.20), the standard deviation σ_s of the assumed *Gaussian* distribution of the barrier heights was obtained from the plot of the effective or apparent zero-bias barrier heights versus $1/T$ which is shown in Fig. 2.13b. The least-squares fit to the *modified Richardson* plot in Fig. 2.13a gives $\Phi_{Bn}^{hom}(0)$ and A_R^* as 0.93 eV and 138 A cm^{-2} K^{-2}, respectively. The same value of $\Phi_{Bn}^{hom}(0)$ is obtained from the least-squares fit in Fig. 2.13b. The effective *Richardson* constant is only slightly larger than the theoretical value, 120 A cm^{-2} K^{-2}.

It also follows from the approach of Song et al. (1986) and Werner and Güttler (1991) that the temperature coefficient $\beta = \partial\Phi_{Bn}/\partial e_0 V_c = 1 - 1/n$ of the effective barrier height of *Schottky* contacts, as defined in Eq. (2.4), varies directly proportional to the inverse temperature. Figure 2.14a proves this prediction. Finally, Fig. 2.14b also shows that the

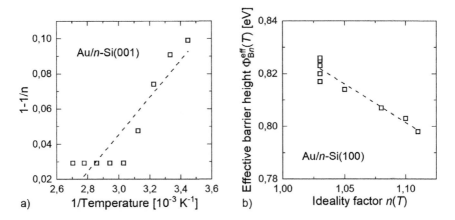

Fig. 2.14 Ideality factors as a function of the reciprocal temperature (**a**) and effective barrier heights as a function of the ideality factors of the same Au/n-Si(001) contact considered in Fig. 2.13. Data from Mahato and Puigdollers (2018)

effective barrier heights and ideality factors evaluated by Mahato and Puigdollers (2018) are linearly correlated as to be expected from relation (2.16) (Schmitsdorf et al., 1995). The extrapolation of the linear least-squares fit gives the barrier height of the laterally homogeneous contacts as 0.83 ± 0.05 eV. Even within the margins of error this value is smaller than the result obtained above from the plots displayed in Fig. 2.13a and b. However, it is worth mentioning, that the barrier height Φ_{Bn}^{hom} obtained from the Φ_{Bn}^{eff}-versus-n plot of Fig. 2.14b excellently agrees with the center of the *Gaussian* distribution of the BEEM barrier heights shown in Fig. 2.12. Obviously, relation (2.16) applies not only to data collected with many *Schottky* contacts of the same type at the same temperature, but also to data evaluated with one contact at different temperatures.

2.5   Extrinsic Interface Dipoles

2.5.1   The Si(111)–(7 × 7)^i Interface Reconstruction

The common feature of reconstructions with large unit meshes on semiconductor surfaces is the reduction of the total surface energy by decreasing the number of dangling bonds per unit area. Figure 2.15 shows the Dimer-Adatom-Stacking Fault (DAS) model of the unit mesh of the Si(111)–(7 × 7) surface reconstruction as it has finally emerged from many experiments and extensive discussions (Takayanagi et al., 1985). It consists of

- a stacking fault in one of its triangular halves,
- a corner hole,

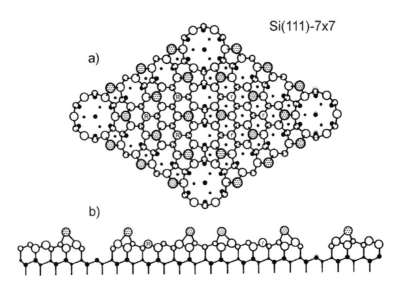

Fig. 2.15 Atom arrangement on clean Si(111)-7 × 7 surfaces according to the dimer-adatom-stacking fault (DAS) model; **a** top view, **b** side view. Adatoms are shaded; restatoms in the faulted and non-faulted triangles of the 7 × 7 unit mesh are marked R and r, respectively. After Takayanagi et al. (1985)

- 9 dimers along the boundary of the faulted, triangular subunit,
- 12 adatoms in T_4 sites in a 2 × 2-like arrangement and, as a consequence of this,
- 6 restatoms between the adatoms in the layer below them.

The Si(111)-7 × 7 reconstruction thus reduces the number of dangling bonds from 49 per 7 × 7 unit cells on an ideally terminated (1 × 1) surface to 19 per one 7 × 7 unit mesh, i.e., by a factor of 3.06.

The geometric structure of the Si(111)–(7 × 7)^i interface reconstruction was determined with Ag- and Pb/Si(111) *Schottky* contacts by grazing-incidence X-ray diffraction (Hong et al. 1992, Grey et al. 1989). The adatoms were found to be missing, while all other elements of the clean-surface structure persist.

A characteristic feature of surface and interface reconstructions is that surface and interface atoms of the substrate are shifted compared to the atomic positions in the corresponding bulk planes. Such structural rearrangements are associated with redistributions of the valence charge. The bonds in perfectly ordered bulk silicon are purely covalent, so the reconstructions are accompanied by $Si^{-\Delta q}$-$Si^{+\Delta q}$ dipoles. The shifted surface components of the Si(2p) core levels observed with (7 × 7)-reconstructed surfaces prove the existence of reconstruction-induced charge transfer.[3]

---

[3] For more details the reader is referred to Sect. 11.4.2 of the monograph by Mönch (2001) and references cited therein.

Chou et al. (1985) calculated the electronic structure of stacking faults in bulk silicon. The left chart of Fig. 2.16 shows the integrated charge-density difference across (111)-planes between a Si crystal with and without an extrinsic stacking fault (ESF) which is schematically illustrated on the right side of the figure. In the ESF, the stacking sequence ...AA'BB'CC'AA'BB'CC'... of the diamond structure is changed to ...AA'BB'AA'CC'AA'BB'.... The calculated charge density difference can be described by two symmetrically arranged electric double layers. Figures 2.15 and 2.16 show that the stacking fault of the Si(111)-7 × 7 surface structure, and the same also holds for the Si(111)–(7 × 7)^i interface structure, corresponds to the lower half of a stacking fault in the bulk. Thus, an electric double layer Si$^{-\Delta q}$ – Si$^{+\Delta q}$ is induced by the stacking-fault, and its negatively charged layer Si$^{-\Delta q}$ is located on the metal side of Si(111)–(7 × 7)^i *Schottky* contacts. The stacking fault of the Si(111)–(7 × 7)^i interface structure causes a decrease of the barrier height of *n*-Si *Schottky* contacts compared to those with the ideal (1 × 1)^i interface structure. Numerical integration of the calculated charge density, displayed in Fig. 2.16, yields an estimated decrease by 37 meV. This estimate is close to the experimental value of –47 meV obtained from the *I/V* characteristic of the Ag/*n*-Si(111)–(1 × 1)^i and -(7 × 7)^i contacts in Fig. 2.7 (Schmitsdorf et al. 1995). As will be discussed in Sect. 2.5.2, the effect of the (7 × 7)^i stacking-fault dipole is completely analogous to the influence of an interlayer of H atoms in metal/H/*p*-Si(111) and -(001) and metal/H/*p*-diamond contacts.

Some metal-silicides can be epitaxially grown on silicon substrates with different interface structures. The two best studied cases are NiSi$_2$/Si(111) and CoSi$_2$/Si(001). Both silicides crystallize in the cubic CaF$_2$ structure. Here, CoSi$_2$/*n*-Si(001) *Schottky* contacts are considered as an example. Werner et al. (1993) determined the atomic interface structure of two different epitaxial structures by transmission electron microscopy and found differently co-ordinated Co interface atoms. They also measured the *I/V* characteristics of contacts with both interface structures. The respective Φ_{Bn}^{eff}-versus-*n* plots of the two types of contacts are displayed in Fig. 2.17. As shown by the data of the Ag/*n*-Si(111)–(7 × 7)^i and -(1 × 1)^i contacts in Fig. 2.7, the effective barrier heights and ideality factors of both types of CoSi$_2$/*n*-Si(001) contacts vary in a linearly correlated manner. The offset of the barrier heights extrapolated to *n* = 1 caused by the differently coordinated Co interface atoms amounts 140 meV and is much larger than the 47 meV attributed to the stacking fault of the (7 × 7)^I interface reconstruction reported by Schmitsdorf et al. (1995). Werner et al. (1993) did not further analyze the correlation between the interface structures of the two CoSi$_2$/*n*-Si(001) contacts and their barrier heights and ideality factors, which they had observed and reported.

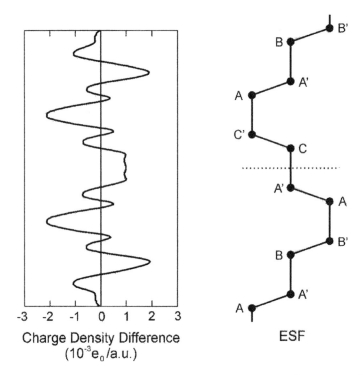

Charge Density Difference
$(10^{-3}\,e_0\,/a.u.)$ ESF

Fig. 2.16 Integrated difference of the charge density over (111) planes between a silicon crystal with an extrinsic stacking fault (ESF) and a perfect silicon crystal, and atomic positions in the (110) plane of a silicon crystal with an extrinsic stacking fault (1 a.u. $= a_B^3 = 1.48 \times 10^{-4}$ nm^3, where a_B is the Bohr radius). After Chou et al. (1985)

Fig. 2.17 Effective barrier heights versus idealty factors of CoSi2/*n-Si*(001) contacts with two different co-ordinations of the Co interface atoms. The dashed and dash-dotted lines are linear least-squares fits to the data. Data from Werner et al. (1993)

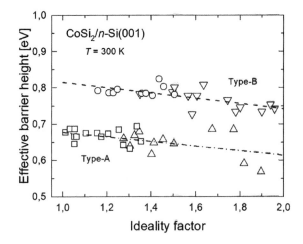

2.5.2 H-modified *Schottky* Contacts

There is much interest in modifying the electronic properties of *Schottky* contacts by interlayers. The simplest but most informative examples are H interlayers in Si and diamond *Schottky* contacts. Figure 2.18,[4,5] displays experimental barrier heights of *p*-Si and *p*-diamond *Schottky* contacts fabricated either on "clean" surfaces or on surfaces precovered with hydrogen. The two semiconductors apparently show opposite behavior in that the barrier heights of *p*-Si and *p*-diamond *Schottky* contacts modified with H interlayers are by 0.16 or 0.30 eV larger, depending on the substrate orientation (111) or (001), and by 1.4 eV smaller, respectively, than the values of the contacts with "clean" interfaces. These observations are easily explained by applying Pauling's electronegativity concept (Mönch, 1994b). The *Pauling* electronegativity of H, 2.20, is larger but smaller than the corresponding value of silicon, 1.90, and of carbon, 2.55, respectively. The bond dipoles between the hydrogen and the semiconductor atoms at the interfaces thus have opposite sign, i.e., metal-H$^{-\Delta q}$ – Si$^{+\Delta q}$ – Si– .... but metal-H$^{+\Delta q}$ – C$^{-\Delta q}$ – diamond configurations are predicted. This conclusion is in complete agreement with the explanation of the lower barrier heights of *n*-Si(111)–(7 × 7)^{i} *Schottky* contacts compared with the values observed with *n*-Si(111)–(1 × 1)^{i} *Schottky* contacts discussed in Sect. 2.5.1. The charge distribution calculated for stacking fault characteristic of the Si(111)–(7 × 7)^{i} reconstruction resulted in the configuration metal- Si$^{-\Delta q}$ – Si$^{+\Delta q}$ – Si– ..., i.e., the orientations of the extrinsic dipoles are the same at the interfaces of H-Si(111), H-Si(001) and Si(111)–(7 × 7)^{I} *Schottky* contacts. Mönch (1994b, 2004) described the extrinsic H-induced interface dipole layers by a simple parallel-plate capacitor and estimated the respective barrier height variations. He obtained the H-induced changes of the barrier heights $\delta\Phi_{Bp}^{H-Si} = +1.00/\varepsilon_i$ eV and $\delta\Phi_{Bp}^{H-C} = -1.86/\varepsilon_i$ eV where ε_i is the respective interface dielectric constant. Simple models give ε_i (H-Si) $\approx$ 4.6 and ε_i (H-C) $\approx$ 2 (Mönch 2004). In any case, the H-induced changes in barrier heights calculated with these rough estimates confirm the magnitude of the experimental results.[6]

---

[4] Figure 2.18a: Experimental data from Aydin et al. (2004), Cetin et al. (2004), Connelly et al. (2004, 2006), Hagemann et al. (2018), Huba et al. (2009), Kampen and Mönch (1995), Karatas et al. (2003), Krix et al. (2009), Nienhaus et al. (2007), Nuhoglu et al. (2003), Zaima et al. (1995).

[5] Figure 2.18b: Experimental data from Aoki and Kawarada (1994), Kawarada et al. (1994), Majdi et al. (2010), Mead and McGill (1976), Ueda et al. (2013).

[6] In a series of articles, Tung and coworkers (Li et al., 2011; Long et al., 2013, 2014) fabricated and studied Ag/*n,p*-Si *Schottky* contacts with interlayers of various non-metal and metal atoms. They interpreted their experimental results in the same way as Mönch (1994), but without mentioning this work and the respective earlier experimental studies discussed in Sect. 2.5.2.

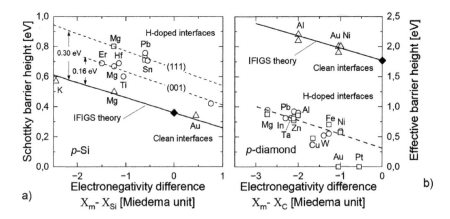

Fig. 2.18 Barrier heights of *p*-Si (**a**) and *p*-diamond (**b**) *Schottky* contacts with clean and H-doped interfaces. The ◆ data points mark the respective *p*-type branch-point energies and the full lines are the predictions of the IFIGS theory (see Sect. 4.1.1). The dashed lines are the linear least-squares fits to the experimental data of the *Schottky* contacts with H-doped interfaces

2.6   Specific Experimental Results

Before turning to the third mechanism inferred from the Φ_{Bn}^{eff} versus *n* diagrams of the Ag/*n*-Si(111)- $(1 \times 1)^i$ and -$(7 \times 7)^i$ contacts shown in Fig. 2.7, namely *the* intrinsic mechanism that explains the barrier heights of ideal *Schottky* contacts, some other important experimental facts should be considered.

2.6.1   Crystallographic Orientation of the Semiconductor

As a first example, some experimental results obtained with well-prepared Au/*n*-Si *Schottky* contact will be considered. In Fig. 2.19a, the effective barrier heights $\Phi_{Bn}^{eff}(T)$ of an Au/*n*-Si(111) and of two Au/*n*-Si(001) contacts are plotted as a function of their ideality factors *n(T)* which were evaluated from their *I/V* characteristics recorded in the temperature range between 100 and 350 K. Extrapolations of the linear fits of the experimental data to *n* = 1 give barrier heights of 0.83, 0.85, and 0.86 eV, respectively, which agree within the error margins of ±0.07 eV. Figure 2.19b, on the other hand, shows the effective barrier heights $\Phi_{Bn}^{eff}(T)$ of one of the Au/*n*-Si(001) contacts considered in Fig. 2.19a and the barrier heights $\Phi_{Bn}^{eff}(300\ K)$ of 23 identically fabricated Au/*n*-Si(001) contacts studied at room temperature as a function of the respective ideality factors. These 32 data points can be jointly described by one linear least-squares fit which then gives an extrapolated barrier height of 0.83±0.06 eV. The above analysis of experimental data

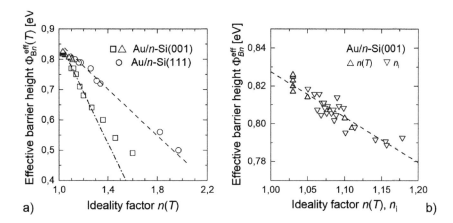

Fig. 2.19 Effective barrier heights versus ideality factors of Au/n-Si(001) and Au/n-Si(111) contacts. The corresponding I/V curves were recorded in the temperature range between 100 and 350 K. The △ data of one of the Au/n-Si(001) diodes are the same as in Fig. 2.14b. The dashed and dash-dotted lines are linear least-squares fits to the data. Data from Maeda and Kitahara (1998), Chen et al. (1993), Sağlam et al. (2004), and Mahato and Puigdollers (2018)

obtained by four different research groups with their well-prepared Au/n-Si *Schottky* contacts leads to the following conclusions. First, the barrier heights of laterally homogeneous *Schottky* contacts can be reliably determined from Φ_{Bn}^{eff} versus n diagrams obtained either with one contact studied at different temperatures or with many contacts prepared under identical conditions and studied at the same temperature. Second, the experimental data clearly demonstrate that the barrier heights of laterally homogeneous *Schottky* contacts are identical whether the Si substrates are (111)- or (001)-oriented. The same conclusion was reached by Nishimura et al. (2007) for Al and Au contacts on (111)-, (110)-, and (001)-oriented Ge substrates. This means that the barrier heights of laterally homogeneous *Schottky* contacts generally do not depend on the respective crystal orientation of the semiconductor substrates if no epitaxial contacts are considered.

2.6.2 *P*- and *n*-Type *Schottky* Contacts

In the case of silicon, *Schottky* contacts were mostly studied on n-type rather than on p-type substrates, presumably because the former generally have larger barrier heights. Figures 2.20 and 2.21 display experimental barrier heights obtained with silicide contacts on both n- and p-type silicon, and Fig. 2.22 shows experimental data of Au/n,p-InP(001) *Schottky* contacts. The barrier heights of the CrSi$_2$/n,p-Si(001) *Schottky* contacts shown in Fig. 2.21a were obtained by using Internal Photoemission Yield Spectroscopy (IPEYS). This is another technique for determining the barrier heights of the laterally homogeneous regions of *Schottky* contacts (for more details see, for example, Sect. 3.8 of the monograph by Mönch (2004)). The common observation of the experimental data plotted in Figs. 2.20, 2.21 and 2.22 is that the barrier heights of laterally homogenous *n*- and *p*-type *Schottky* contacts of the same kind add up to the band gap of the respective semiconductor, as shown by the data summarized in Table 2.1.

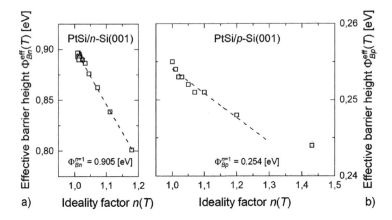

Fig. 2.20 Effective barrier heights of one PtSi/*n*-Si(001) (**a**) and one PtSi/*p*-Si(001) contact (**b**) as a function of the ideality factors in the temperature range between 100 and 300 K. Data from Wittmer (1991) and Chin et al. (1993), respectively

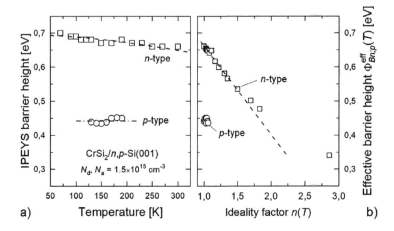

Fig. 2.21 Temperature dependence of IPEYS barrier heights (**a**) and effective barrier heights versus ideality factors (**b**) of CrSi$_2$ contacts on one Si(001) substrate doped *n*- and *p*-type each. The dashed lines are explained in the text. Data from Aniltürk and Turan (1999)

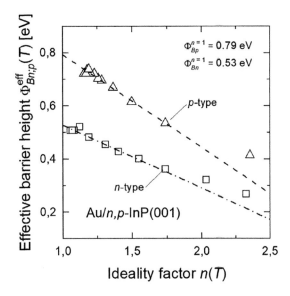

Fig. 2.22 Effective barrier heights of Au/*n,p*-InP(001) *Schottky* contacts as a function of the ideality factors in the temperature range from 80 to 320 K. Data from Horvath et al. (2005)

Table 2.1 Barrier heights Φ_{Bn}^{hom} and Φ_{Bp}^{hom} of laterally homogenous *n*- and *p*-type *Schottky* contacts, respectively, from the data plotted in Figs. 2.20, 2.21 and 2.22 and band gaps W_g of Si and InP, respectively

	Φ_{Bn}^{hom} [eV]	Φ_{Bp}^{hom} [eV]	$\Phi_{Bn}^{hom} + \Phi_{Bp}^{hom}$ [eV]	W_g [eV]
PtSi/Si(001)	0.90	0.25	1.15	1.12
CrSi$_2$/Si(001)	0.66	0.44	1.10	1.12
Au/InP(001)	0.53	0.79	1.32	1.34

The experimental data plotted in Fig. 2.21a reveal another interesting and important observation that applies to all *Schottky* contacts: the temperature dependence of the barrier heights of laterally homogeneous *n*-type *Schottky* contacts is the same as that of the fundamental bulk band gap of the respective semiconductor, while the barrier heights of laterally homogeneous *p*-type *Schottky* contacts are (almost) independent of temperature. The dashed line in Fig. 2.21a represents the temperature variation of the fundamental bulk band gap of silicon.

2.6.3 *Schottky* Contacts of SiC Polytypes

Silicon carbide crystallizes in a variety of polytypes characterized by different band gaps. Figure 2.23 displays barrier heights of Pd *Schottky* contacts on (0001)-oriented samples of the two hexagonal 4*H* and 6*H* polytypes, which have band gaps of 3.23 and 3.0 eV,

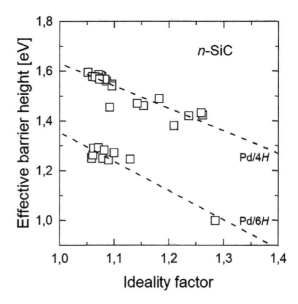

Fig. 2.23 Barrier heights of Pd/n-$4H$-SiC(0001) and Pd/n-$6H$-SiC(0001) contacts as a function of the ideality factors at room temperature. The dashed lines are linear least-squares fits to the experimental data. Data from Im et al. (1998)

respectively. The barrier height extrapolated to $n = 1$ measure 1.63 and 1.35 eV for the $4H$ and the $6H$ sample, respectively. Within the margins of error, the difference of 0.28 eV agrees with the difference of 0.23 eV of the respective band gaps. Taking into account the observation just mentioned that the barrier heights of n- and p-type *Schottky* contacts add to the band gap, the p-type *Schottky* barrier heights of the $4H$- and $6H$-SiC polytypes are the same. This had been already proposed by Mönch (1994a).

From the *Schottky–Mott* Rule to Interface-Induced Gap States

3

3.1   *Schottky–Mott* Rule

Mott (1938) was the first to propose a most simple rule for predicting the barrier heights of metal–semiconductor contacts. In a thought experiment, he considered a metal and an *n*-type semiconductor that has no surface states in the band gap and thus flat bands up to the surface that face each other in a vacuum. Since the work functions of the metal and the semiconductor are generally different, an electric field will exist in the vacuum gap between them. Therefore, both the metal and the semiconductor carry surface charges of the same density, Q_m and Q_s, respectively, but opposite sign. The condition of charge neutrality is

$$Q_m + Q_s = Q_m + Q_{sc} = 0. \tag{3.1}$$

The penetration lengths of the electric field into the two opposing solids are very different. Because of the large electron densities in metals, the *Thomas–Fermi* screening lengths usually measure less than a tenth of a nanometer. Thus, electric field penetration at metal surfaces can be neglected. Screening in a semiconductor that is non-degenerately doped is described by its *Debye* length which, for example, amounts to 13 nm for a doping level of 10^{17} cm^{-3} at room temperature. Therefore, electric fields penetrate quite deeply into non-degenerate semiconductors, resulting in extended space-charge layers. In the example of Fig. 3.1, the metal is assumed to have a larger work function than the non-degenerately doped *n*-type semiconductor. Therefore, the surface space-charge Q_{sc} of the semiconductor has positive sign, i.e., the semiconductor surface is depleted of mobile electrons and the bands are bend upwards. When the distance between the metal and the semiconductor is reduced, the strength of the electric field between them decreases and the surface band-bending of the semiconductor increases accordingly. In the *mathematical*

© The Author(s), under exclusive license to Springer Nature Switzerland AG 2024   37
W. Mönch, *Electronic Structure of Semiconductor Interfaces*, Synthesis Lectures on Engineering, Science, and Technology, https://doi.org/10.1007/978-3-031-59064-1_3

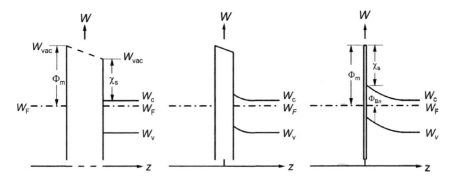

Fig. 3.1 Development of a *Schottky* barrier as a function of decreasing metal-to-semiconductor distance. The semiconductor and the interface are assumed to be free of surface and interface states, respectively

limit of an intimate contact, the vacuum levels of the metal and the semiconductor finally align and the barrier height of the Schottky contact

$$\Phi_{Bn}^{S-M} = W_{ci} - W_F = \Phi_m - \chi_s. \tag{3.2}$$

equals the difference between the metal work-function Φ_m and the electron affinity $\chi_s = W_{vac} - W_{cs}$ of the semiconductor. This is the famous *Schottky–Mott* rule, as it was called later. When the semiconductor is doped p-type the barrier height of holes

$$\Phi_{Bp}^{S-M} = W_F - W_{vi} = I - \Phi_m \tag{3.3}$$

equals the difference between the ionization energy $I = W_{vac} - W_{vs}$ of the semiconductor and the metal work-function.

Already in 1939, Schweikert performed the first test of the *Schottky–Mott* rule. He measured the *I/V* characteristics of various metal-selenium rectifiers and determined their initial resistance. The early diffusion theory of Schottky and Spenke (1939) as well as Bethe's (1942) later thermionic emission theory revealed the zero-bias resistance of *Schottky* contacts to depend exponentially on the barrier height. Figure 3.2 displays Schweickert's data (1939) as published by Schottky (1940). Several basic conclusions were drawn from these experimental results. First, the observed decrease of the barrier heights with increasing metal work-function proved that, according to relation (3.3), selenium is a p-type semiconductor. Second, as predicted by the *Schottky–Mott* rule, the barrier heights of the metal-Se contacts indeed vary linearly as a function of the metal work-function but the slope parameter $S_\Phi = -d\Phi_{Bp}/d\Phi_m$ is only 0.08 instead of 1 as demanded by the *Schottky–Mott* rule (3.2).

Already then, it was generally accepted that the simple *Schottky–Mott* rule was not supported by the experimental results of real metal–semiconductor contacts. This common

Fig. 3.2 Zero-bias resistance of selenium *Schottky* contacts as a function of the metal work-function. The dashed line is a linear least-squares fit to the data, which gives a slope parameter $S_X = -0.08$. Data from Schweikert (1939) as reported by Schottky (1940). From Mönch (1986)

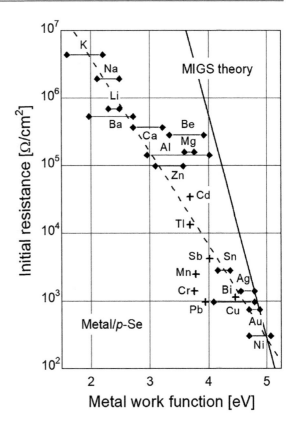

view was also strengthened by Mott's most important constraint under which he derived his rule, namely *"the two* (i.e., the metal and the semiconductor) *do not influence each other in any way"*. Therefore, neither Shive (1940) nor Brattain (1940) nor Welker (1943), who made similar observations with Se, Cu$_2$O, and Ge Schottky contacts, respectively, published their data. Therefore, I am surprised that even now, in 2024, the *Schottky–Mott* rule is still invoked in many articles to explain barrier heights of *Schottky* contacts.

3.2   Bardeen's Solution: Interface States

Already a few years later in 1947, Bardeen fundamentally solved this dilemma of the slope parameter. He proposed that interface states on the semiconductor side of *Schottky* contacts could absorb charge in addition to the space-charge layer. The condition of charge neutrality now reads

$$Q_m + Q_s = Q_m + Q_{sc} + Q_{is} = 0, \tag{3.4}$$

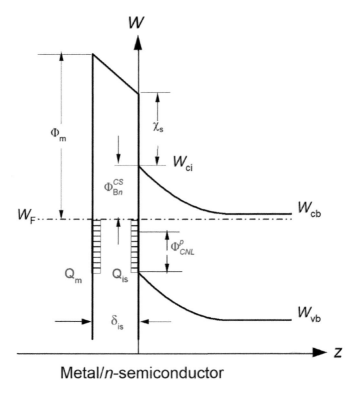

Fig. 3.3 Energy band diagram of a *Schottky* contact with interface states. After Cowley and Sze (1965)

where Q_{is} is the charge density in the interface states. For the two limiting cases of very large and almost vanishing densities of interface states, their influence is obvious. In the latter case, the interface states will have no significant effect, so that the *Schottky–Mott* rule with $S_\Phi \approx 1$ is retrieved. On the other hand, if the density of interface states is very large, they will absorb most of the charge on the semiconductor side of the contact. Consequently, different charge densities on the metal side of the contact will then lead to only small variations of the space-charge density and hence the barrier height. The slope parameter will then be small, i.e., $S_\Phi \ll 1$. Such behavior was called the *Bardeen* case.

3.3 *Cowley-Sze* Modell: *Schottky–Mott* Modell Plus Interface States

Cowley and Sze (1965) adopted Bardeen's proposal and considered an n-type semiconductor which they assumed to have a continuum of surface states with a constant density of states D_{is} across the band gap. Above and below their charge-neutrality level W_{CNL} these states originate from the conduction and the valence band, respectively. To form a *Schottky* contact in Mott's Gedanken experiment, Cowley and Sze approached the metal and the semiconductor up to a separation δ_{is} of a few angstroms. They assumed that this interfacial layer is transparent to electrons whose energies are greater than the band bending at the interface. Consequently, they now defined the initial *surface* states as *interface* states. Finally, they obtained the barrier height of their *Schottky* contact as

$$\Phi_{Bn}^{C-S} = S_\Phi(\Phi_m - \chi_S) + (1 - S_\Phi)\left(W_g - \Phi_{CNL}^p\right)$$
$$= S_\Phi(\Phi_m - \chi_S) + (1 - S_\Phi)\Phi_{CNL}^n \quad (3.5)$$

and the slope parameter as

$$S_\Phi = 1/(1 + e_0\delta_{is}D_{is}/\varepsilon_0\varepsilon_i), \quad (3.6)$$

where $W_g = W_c - W_v$ is the band gap of the semiconductor, $\Phi_{CNL}^p = W_{CNL} - W_v$ and $\Phi_{CNL}^n = W_g - \Phi_{CNL}^p = W_c - W_{CNL}$ are the p-type and n-type CNL energies, respectively, and ε_i is the dielectric constant of the interfacial layer. Rearranging the terms of relation (3.5) yields

$$\Phi_{Bn}^{C-S} = \Phi_{CNL}^n + S_\Phi\left(\Phi_m - \chi_S - \Phi_{CNL}^n\right)$$
$$= \Phi_{CNL}^n + S_\Phi(\Phi_m - W_{vac} + W_C - W_C + W_{CNL})$$
$$= \Phi_{CNL}^n + S_\Phi(\Phi_m - \Phi_{CNL}) \quad (3.7)$$

The interfacial layer of width δ_{is}, as assumed by Cowley and Sze (1965), represents an electric dipole layer given by the difference of the work function Φ_m of the metal and the CNL work function $\Phi_{CNL} = W_{vac} - W_{CNL}$ of the semiconductor. This means that, as in the *Schottky–Mott* rule (3.3, 3.4), the essential interface dipole is again expressed by two *surface* properties.

3.4 Critique of the Use of Surface Properties

The work functions Φ_m of metals differ by up to 0.6 eV depending on the surface orientation; see, for example, the compilation by Michaelson (1977). Therefore, the work functions of polycrystalline metals are usually considered in the analysis of the barrier

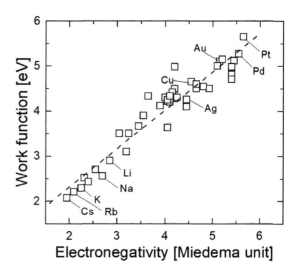

Fig. 3.4 Work functions of polycrystalline metals as a function of Miedema's electronegativities. Work functions from Michaelson (1977)

heights of *Schottky* contacts. However, neither the distributions of the grain orientations nor their dimensions are specified.

Already in 1956, Gordy and Thomas found that the work functions of metals are linearly related to their electronegativities. The linear least-squares fits to the data plotted in Fig. 3.4 are

$$\Phi_m = 0.86 \mathrm{X}_{Mied} + 0.59 \text{ [eV]} \tag{3.8}$$

if Miedema's (1973, 1980) electronegativities are used and

$$\Phi_m = 1.79 \mathrm{X}_{Paul} + 1.11 \text{ [eV]} \tag{3.9}$$

if Pauling's (1939/1960) electronegativities are applied instead. Miedema (1973, 1980) derived his electronegativity scale from properties of bulk metals, i.e., solids. In this article, his scale will be used whenever properties of metals and semiconductors are discussed. Pauling, on the other hand, obtained his values from the analysis of molecular dipoles. In analogy to molecules (Pauling, 1939/1960), the average electronegativities of compounds are assumed to be the geometric means of the atomic values of their constituents.

Based on the findings of Gordy and Thomas (1956), Mead (1966) then argued with respect to *Schottky* contacts that *"the work function may be thought of as composed of two parts. One is related to the chemical electronegativity of the metal and the other is a surface dipole layer caused by a distortion of the electronic cloud at the surface. The surface dipole layer is certainly not present in anything like the same form when the metal is in intimate contact with a semiconductor as it is when the metal faces a vacuum"*. Therefore, he suggested to "*use the electronegativity of the metal as its reference property rather than the work function*".

Of course, Mead's (1966) arguments above can also be applied to semiconductors. In terms of semiconductor interfaces, this then means that the electronegativities of semiconductors should be used instead of their electron affinities or ionization energies.

3.5 Formation of *Schottky* Contacts

3.5.1 Adatom-Induced Surface States and Dipoles

Mead's well-founded suggestion is still, as of December 2023, ignored in most discussions of barrier heights of *Schottky* contacts with different metals on the same semiconductor. The most common argument against the use of electronegativities is that they are *conceptual* quantities, but not precisely defined and measurable or calculable from first principles. However, as will be shown below, there are other well-reasoned arguments based on experimental data to characterize the electronic properties of semiconductor interfaces by electronegativities rather than surface properties. To this end, we will follow the formation of *Schottky* contacts by first considering the first monolayer of metal adatoms deposited stepwise on clean semiconductor surfaces. Silicon surfaces covered with cesium is considered as an example.

Figure 3.5 shows the work function $\Phi_s(\Theta)$ of clean, cleaved Si(111) surfaces as a function of Cs coverage Θ as measured by Allen and Gobeli (1966). The observed decrease $\Delta\Phi_s$ cannot simply be attributed to the formation of adatom-induced $Cs^{+\delta q}-Si^{-\delta q}$ dipoles, in analogy to Cs-covered metal surfaces.

For semiconductors, in addition to the adatom-induced surface dipoles, which cause variations $\Delta I = \Delta(W_{vac} - W_{vs})$ of the ionization energy, possible adatom-induced changes $\Delta(W_F - W_{vs})$ in the surface band bending must be taken into account. The changes in the work-function of semiconductors can be reasonably written as

$$\Delta\Phi_S = \Delta(W_{vac} - W_F)$$
$$= \Delta(W_{vac} - W_{vs}) - \Delta(W_F - W_{vs})$$
$$= \Delta I - \Delta(W_F - W_{vs}). \tag{3.10}$$

The changes $\Delta(W_F - W_{vs})$ in the position of the *Fermi* level with respect to the valence-band maximum indicate modifications in the distribution of surface states in the band gap caused by the adatoms. Therefore, a complete analysis of adatom-induced changes of the work function of semiconductors requires a second, independent experimental approach.

Figure 3.6 displays the experimentally determined position $W_F - W_{vs}$ of the *Fermi* level above the valence-band maximum on initially clean, cleaved Si(111) surfaces in dependence of the Cs coverage (Mönch, 1970). Initially, the *Fermi* level is pinned 0.36 eV

Fig. 3.5 Work function of clean, cleaved Si(111)-2 × 1 surfaces as a function of Cs coverage. Data from Allen and Gobeli (1966)

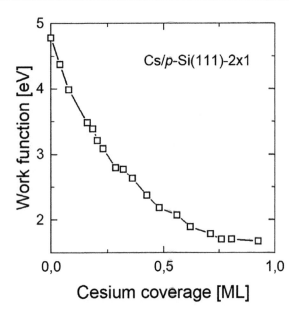

above the valence-band maximum by the dangling-bond surface states characteristic of the clean, cleaved Si(111) surface. With increasing Cs coverage, the *Fermi* level shifts to a new pinning position closer to the conduction-band minimum, indicating the formation of Cs-induced surfaces states. The dashed curve was calculated for a coverage-proportional decrease and increase of the clean-surface and the Cs-induced surface states, respectively.

Substituting the experimental data shown in Figs. 3.5 and 3.6 into Eq. (3.10), one obtains the decrease in the ionization energy plotted in Fig. 3.7 as a function of the Cs coverage. Topping (1927) attributed the non-linear reduction of the ionization energy for larger Cs coverages to a depolarizing interaction of the adatom-induced surface dipoles oriented normal to the surface and parallel to each other. He obtained the variation of the ionization energy as a function of adatom coverage Θ as

$$\Delta I = \pm \frac{e_0 p_0 \sigma \Theta}{\varepsilon_0} \frac{1}{9\alpha_{ad} \sigma^{3/2} \Theta^{3/2}}, \tag{3.11}$$

where p_0 is the dipole moment of an isolated adatom-induced surface dipole, σ is the number of surface sites per monolayer, and α_{ad} is the polarizability of an adatom. The dashed curve in Fig. 3.7 is a least-squares fit to the experimental data with $p_0 = 1.3 \times 10^{-27}$ As cm and $\alpha_{Cs} = 16.3 \times 10^{-24} \text{cm}^{-3}$.

The dipole moment of Cs-Si surface bonds can be estimated by applying Pauling's (1939/1960) electronegativity concept. He correlated the ionicity δq_1 of single bonds in diatomic molecules $A - B$ with the difference $X_A - X_B$ of the atomic electronegativities of the two different atoms covalently bonded. Here, the revised relation derived by Hanney and Smith (1946)

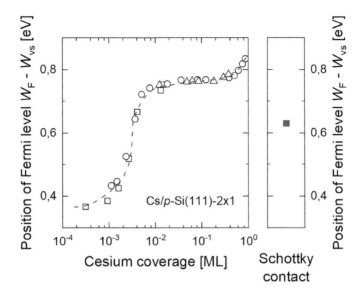

Fig. 3.6 Position of Fermi level with respect to the valence-band maximum of a clean, cleaved *p*-Si(111)-2 × 1 surface as a function of Cs coverage and of a Cs/*n*-Si(111)-7 × 7 Schottky contact. Data from Mönch (1970) and Weyers et al. (1999), respectively

Fig. 3.7 Decrease of the ionization energy of a clean, cleaved Si(111)-2 × 1 surface as a function of Cs coverage. From Mönch (1970)

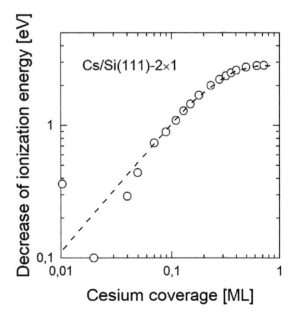

Table 3.1 Electronegativities and covalent radii of Cs and Si

	X_A [Pauling units]	r_{cov} [nm]
Cs	0.79	0.235
Si	1.9	0.117

$$\delta q_1 = 0.16|X_A - X_B| + 0.035|X_A - X_B|^2 \tag{3.12}$$

will be applied. Using the simple point-charge model, the dipole moment of a covalently bonded A–B molecule is then given by

$$p_{A-B} = \delta q_1 e_0 d_{cov} = e_0 \left(0.16|X_A - X_B| + 0.035|X_A - X_B|^2\right)\left(r_{cov}^A + r_{cov}^B\right) \tag{3.13}$$

where r_{cov}^A and r_{cov}^B are the covalent radii of the A and B atoms, respectively. Using the data listed in Table 3.1, the dipole moment of Cs $-$ Si surface molecules is found to be $p_{Cs-Si} = 1.47 \times 10^{-27}$ As cm. This estimate is close to the above result obtained from the least-squares fit to the experimental Cs-induced decrease of the ionization energy of cleaved Si surfaces shown in Fig. 3.7.

3.5.2 Adatom-Induced Surface Core-Level Shifts

The adatom-induced charge transfer δq_1 changes the electrostatic potential experienced by the core electrons of both adatoms and surface atoms. This leads to chemical shifts of the core-level energies. Silicon surfaces covered with cesium will be again considered as a first example. The Cs-induced shift of the Si(2p) core levels on both Si(111) and Si(001) surfaces has been determined to be −0.57 eV (Lin et al., 1991; Magnussen et al., 1991; Park et al., 1995; Chao et al., 1996). The negative sign of the Si(2p) core-level shift confirms the formation of Cs$^{+\delta q}$ − Si$^{-\delta q}$ already concluded from the observed decrease of the ionization energy.

Even beyond a monolayer coverage, covalent bonds between metal and semiconductor atoms persist at metal–semiconductor interfaces. Zhang et al. (1986) studied the structural and electronic properties of Al/GaAs(110) interfaces. Charge-density contours calculated by them are displayed in Fig. 3.8. They clearly show covalent bonds directly at the interface, with the bond charges shifted somewhat toward both the Ga and As atoms at the interface. The electric dipole moments observed with a monolayer of metal-adatoms on semiconductor surfaces should therefore explain the chemical trends of the barrier heights of the respective *Schottky* contacts.

The barrier heights of laterally homogeneous n-Si(111)-$(1 \times 1)^i$ and $-(7 \times 7)^i$ Schottky contacts are plotted in Fig. 3.9 against the Si(2p) core-level shifts δW_{cl} observed with the respective metal adatoms on Si surfaces. Obviously, the data are linearly correlated in both cases, and the linear least-squares fits yield

Fig. 3.8 Charge contour plots in the $(1\bar{1}0)$ (a and b) and the (100) plane (c) of Al/GaAs(110) contacts. Solid squares are gallium atoms, solid diamonds are arsenic atoms, and aluminum atoms are denoted by solid circles. From Zhang et al. (1986) with kind permission of Prof. M. L. Cohen

$$\Phi_{Bn}^{hom} = (0.80 \pm 0.06) + (0.42 \pm 0.30)\delta W_{cl}(Si_{2p}) \ [eV] \qquad (3.14)$$

for n-Si(111)-(1×1)[i] and

$$\Phi_{Bn}^{hom} = (0.72 \pm 0.04) + (0.42 \pm 0.11)\delta W_{cl}(Si_{2p}) \ [eV] \qquad (3.15)$$

for n-Si(111)-(7×7)[i] Schottky contacts. In contrast to the *Schottky–Mott* rule (3.2) and relation (3.7) of the Cowley-Sze approach, the barrier heights of *Schottky* contacts are now explained by an *interface* property rather than by *surface* properties, namely the partial ionic character of the covalent bonds between metal and semiconductor atoms directly at the interface. However, there are not much experimental data on both the ionization energy changes and the core-level shifts induced by adatoms.

Pauling's electronegativity concept directly addresses the partial ionic character of the covalent bonds. Therefore, Fig. 3.10 shows the experimental shifts of the C(*1s*), Si(2*p*), and Ge(3*d*) core-levels caused by metal-adatoms on diamond, silicon, and germanium substrates, respectively, versus the difference $X_m^{Paul} - X_s^{Paul}$ of the electronegativities of

Fig. 3.9 Barrier heights of laterally homogeneous n-Si(111):(7 × 7)[i] and −(1 × 1)[i] *Schottky* contacts versus the Si(2p) core-level shifts induced by the respective metal adatoms on Si(111) surfaces. From Mönch (2014)

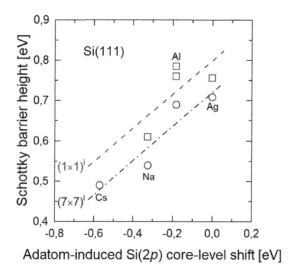

the metals and semiconductors. Obviously, the data are linearly correlated irrespective of the semiconductor, and the linear least-squares fit yields

$$\delta W_{cl} = (0.05 \pm 0.02) + (0.42 \pm 0.02)\left(X_m^{Paul} - X_s^{Paul}\right). \tag{3.16}$$

This behavior is not limited to metal adatoms. The data of nonmetal adatoms on diamond, Si, and Ge show exactly the same dependence; see Fig. 14.21 in the book of Mönch (2001). The adatom-induced core-level shifts observed with both cations and anions of III–V compounds exhibit the same dependence; see Fig. 14.22 in the book of Mönch (2001).

The correlations (3.14) and (3.15) between the experimental barrier heights of *Schottky* contacts and the core-level shifts induced by the respective metal-adatoms, and the linear dependence (3.16) of these core-level shifts on the difference of the metal and the semiconductor electronegativities now justify substituting in Eq. (3.7) the difference $\Phi_m - \Phi_{CNL}$ of the metal work functions and the CNL work-function of the interface states by the difference $X_m - X_s$ of the metal and the semiconductor electronegativities. This means that relation (3.7) is now replaced by

$$\Phi_{Bn}^{hom} = \Phi_{CNL}^n + S_X(X_m - X_s) \tag{3.17}$$

for n-type *Schottky* contacts, and for p-type *Schottky* contacts one obtains

$$\Phi_{Bp}^{hom} = \Phi_{CNL}^p - S_X(X_m - X_s). \tag{3.18}$$

Fig. 3.10 Chemical shifts of C(*1s*), Si(*2p*), and Ge(*3d*) core-levels induced by metal adatoms on diamond, silicon, and germanium surfaces, respectively, as a function of the difference $X_m - X_s$ of the metal and the semiconductor electronegativities in Pauling units. From Mönch (2001) and references cited therein, Tong et al. (2002), Gurnett et al. (2005), and Liu et al. (2013)

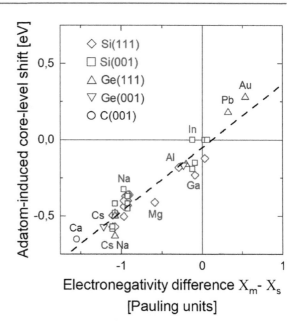

The slope parameter is

$$S_X = d\Phi_m/dX_{Mied} = A_X S_\Phi, \tag{3.19}$$

where, according to relation (3.8), $A_X = 0.86\,\text{eV/Miedemaunit}$. Thus, Mead's (1966) proposal to "*use the electronegativity of the metal as its reference property rather than the work function*" is now finally physically justified based on experimental data.

3.5.3   Electronegativity Concept: Si and InP *Schottky* Contacts

As two examples, the dependence of experimental barrier heights of clean and laterally homogeneous *n*-Si(111)-(1 × 1)[i] and −(7 × 7)[i] and *n*- and *p*-InP Schottky contacts are now discussed as a function of the electronegativity difference $X_m - X_s$. Here and in the following, Miedema's electronegativities are used for the reasons given above. Generalizing Pauling's concept, the electronegativities of compounds are taken as the geometric mean of the atomic values of their constituents.

The data presented in Figs. 3.11 and 3.12 confirm that the barrier heights vary linearly with the electronegativity difference $X_m - X_s$, as expected from relations (3.17) and (3.18). The linear least-squares fit to the data of the Si *Schottky* contacts displayed in Fig. 3.11 yield

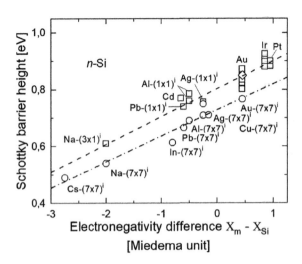

Fig. 3.11 Barrier heights of laterally homogeneous *n*-type silicon *Schottky* contacts versus the difference of the *Miedema* electronegativities of the metal and silicon. The □ and ○ symbols differentiate the data of contacts with Si(111)-(1 × 1)[i]-unreconstructed and −(7 × 7)[i]-reconstructed interfaces, respectively. The dashed and the dash-dotted lines are linear least-squares fits to the □ and ○ data, respectively. From Mönch (2004) and references cited therein, Saglam et al. (2004), Cetin et al. (2004), Akklilic et al. (2004), Rahmatallahpur and Yegane (2011), Long et al. (2013), and Parui et al. (2015)

$$\Phi_{Bn}^{(1 \times 1)i} = (0.803 \pm 0.005) + (0.098 \pm 0.007)(X_m - X_{Si}) \text{ [eV]} \tag{3.20}$$

and

$$\Phi_{Bn}^{(7 \times 7)i} = (0.729 \pm 0.012) + (0.092 \pm 0.008)(X_m - X_{Si}) \text{ [eV]} \tag{3.21}$$

for the *n*-Si(111)-(1 × 1)[i] and −(7 × 7)[i] *Schottky* contacts, respectively. Within the margins of error, the slope parameters are identical. The two values of the zero-charge-transfer energies differ by 73 meV. This difference is the same as observed with the Ag/*n*-Si(111)-(1 × 1)[i] and the −(7 × 7)[i] *Schottky* contacts already discussed in Sects. 2.3 and 2.5.1. It was there attributed to the electric dipole associated with the stacking fault of the Si(111)-(7 × 7)[i] interface structure. Consequently, the fit (3.20) gives the charge neutrality energy $\Phi_{CNL}^n = 0.803\text{eV}$ for *n*-Si(111)-(1 × 1)[I] *Schottky* contacts.

Figure 3.12 shows the barrier heights of laterally homogeneous *n*- and *p*-InP *Schottky* contacts as a function of the difference of the metal and InP electronegativities. The linear least-squares fits to the experimental data yield

$$\Phi_{Bp}^{hom} = (0.88 \pm 0.01) - (0.10 \pm 0.01)(X_m - X_{InP}) \text{ [eV]} \tag{3.22}$$

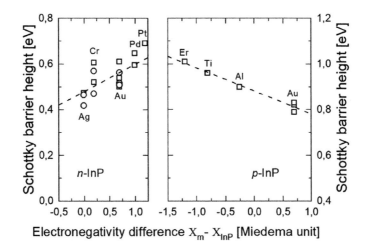

Fig. 3.12 Barrier heights of laterally homogeneous *n*- and *p*-InP *Schottky* contacts as a function of the difference of the metal and InP electronegativities. □ and ○ data are from *I/V* and *C/V* characteristics, respectively. The dashed lines are the linear least-squares fits to the experimental data. From Mönch (2008) and references cited therein, Korucu et al. (2013), Korkut (2013), Kumar et al. (2013), Ejderha et al. (2012), and Chen et al. (2016)

and

$$\Phi_{Bn}^{hom} = (0.48 \pm 0.02) - (0.12 \pm 0.03)(X_m - X_{InP}) \text{ [eV]}, \tag{3.23}$$

respectively. The slope parameters are again identical within the error margins. The two experimental CNL energies $\Phi_{CNL}^{p} = 0.88\text{eV}$ and $\Phi_{CNL}^{n} = 0.48\text{eV}$ add up to 1.36 eV. Within the margins of error, this value corresponds to the width of the InP band gap at room temperature, namely 1.34 eV.

In summary, the two quantities, the CNL energies Φ_{CNL}^{p} and/or Φ_{CNL}^{n} and the slope parameters $S_X = d\Phi_{Bn,p}^{hom}/dX_{Mied}$, that determine the barrier heights of different *Schottky* contacts on the same semiconductor can be obtained from plots of experimental barrier heights of abrupt, intimate, and laterally homogeneous metal–semiconductor contacts as a function of the electronegativity difference $X_m - X_{semi}$ without any further assumptions. This then allows to estimate the barrier heights of other metal–semiconductor combinations.

3.6   Metal-Induced Gap States

The experimental data presented in the previous sections have shown that a continuum of interface states, as assumed by Cowley and Sze (1965), and the description of the charge transfer across metal–semiconductor interfaces by the difference of the electronegativities of metal and semiconductor describe the experimentally observed barrier heights of well-characterized *Schottky* contacts. The only thing missing now is a *physical* model of the interface states which finally allows to calculate their essential parameters, namely their charge neutrality level and density of states as well as the "thickness" of the induced interface-dipole layer. At the same time as Cowley and Sze (1965) presented their *phenomenological approach,* Heine (1965) had independently published a *physical concept* that solves this problem.

At clean metal and semiconductor surfaces the wave-functions of the electrons decay exponentially into vacuum. Figure 3.13a explains this behavior. If, as in a *Schottky* contact, the vacuum is replaced by a semiconductor, the propagation or the wave-functions across the interface is somewhat more complicated. Heine (1965) now argued that in the energy range from the top of the valence band to the *Fermi* level, where the conduction band of the metal overlaps band gap of the semiconductor, the wave-functions of the metal electrons tunnel into the semiconductor, just as they do in vacuum at clean surfaces. This behavior is shown schematically in Fig. 3.13b. As at clean metal surfaces, the tails of the wave-functions of the metal electrons at metal–semiconductor interfaces are described by complex wave vectors. These *metal-induced gap states*, or MIGS for short, as Louie and Cohen (1976) and Louie et al. (1977) later named them, thus derive from the complex band-structure of the semiconductor bulk, i.e., they are an intrinsic property of the semiconductor. The early and simple model calculations of Louie and coworkers (1976, 1977) as well as the results of later, more advanced computations confirmed Heine's original concept perfectly. In the following, this concept is explained considering a simple one-dimensional lattice in the nearly free electron approximation, as first presented by Maue (1935) and later extended by Goodwin (1939).

The periodic potential of a one-dimensional solid can be approximated by keeping only the first *Fourier* coefficient, i.e., by

$$V(z) = V_0 - V_1 \cos(g_1 z), \qquad (3.24)$$

where $g_1 = 2\pi/a$ is the shortest vector of the reciprocal lattice and a is the lattice parameter. Provided that the *Fourier* coefficient V_1 of the potential is small compared to the kinetic energy of the electrons, their wave-functions can be approximated by the first two terms of a *Fourier* expansion, i.e.,

$$\psi_k(z) = A\exp(ikz) + B\exp(i(k - g_1)z. \qquad (3.25)$$

Fig. 3.13 Wavefunctions at
clean surfaces of solids (**a**) and
at metal–semiconductor and
semiconductor–semiconductor
interfaces (**b**) (schematic)

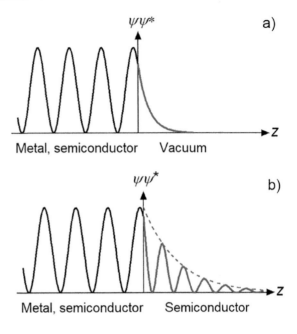

For the following, it is convenient to measure the wave vectors from the edge of the
Brillouin zone, i.e., $k = \pi/a - \kappa$. Substituting (3.24) and (3.25) in Schrödinger's equation,
one obtains the energy dispersion

$$W(\kappa) - V_0 = W_1 + \left(\hbar^2/2m_0\right)\kappa^2 \pm \left[V_1^2 + 4W_1\left(\hbar^2/2m_0\right)\kappa^2\right]^{1/2} \tag{3.26}$$

where the definition

$$W_1 = \left(\hbar^2/2m_0\right)\left(g_1/2\right)^2 \tag{3.27}$$

is used. The left diagram of Fig. 3.14 displays schematically the resulting well-known
band structure for real wave vectors k.[1]

The energy $W(\kappa)$ is a continuous function of κ^2. can assume Both positive and negative values of κ^2, i.e., real and complex wave vectors are equally allowed. In the bulk,
only real wave vectors are physically meaningful because for complex wave vectors the
Bloch waves (3.25) would grow exponentially in the limit of $z \rightarrow \infty$. Thus, they could
not be normalized. At surfaces and interfaces, however, complex wave vectors become
physically relevant. The periodic wave-functions of surface states decay exponentially
from the surface into the bulk of the semiconductor and are fitted to a simple exponential
tail into vacuum, see Fig. 3.15a. Such wave-functions can be easily normalized and thus
represent electrons bound to surfaces.

---

[1] See text books on solid state physics as, for example, Ibach and Lüth (1991).

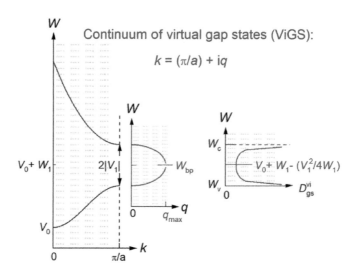

Fig. 3.14 Band structure of a one-dimensional solid for real and complex wave-vectors and density of virtual gap states (schematically)

Surface states, and the same is true for adatom-induced surface states and interface-induced states (see Fig. 3.15a–c), at the end of a finite one-dimensional chain will thus have energies within the band gap at the boundary π/a of the first *Brillouin* zone and complex wave vectors

$$k = \pi/a + iq. \tag{3.28}$$

Analogous to the *Fourier* expansion (3.25), the corresponding wave functions can be written as

$$\psi_q(z) = \exp(-qz)\left[A\exp(i\pi z/a) + B\exp(-i\pi z/a)\right]$$
$$= A'\exp(-qz)\cos(\pi z + \varphi), \tag{3.29}$$

where A' is a constant and the ratio $A/B = \exp(i2\,\varphi)$ of the two *Fourier* coefficient defines the phase factor φ. Its imaginary part is obtained as

$$\sin 2\varphi = -W_1 q/V_1 g_1. \tag{3.30}$$

Across the band gap, φ changes from $-\pi/2$ to 0 and from 0 to $\pi/2$ for positive and negative values of the *Fourier* coefficient V_1 of the potential, respectively.

For complex wave vectors (3.28), the solutions of Schrödinger's equation have the form

$$W(q) - V_0 = W_1 - (\hbar^2/2m_0)q^2 \pm [V_1^2 - 4W_1(\hbar^2/2m_0)q^2]^{1/2} \tag{3.31}$$

Fig. 3.15 Wave-functions of "clean surface" (**a**) and adatom-induced surface states (**b**), and of interface-induced gap states (**c**) (schematic)

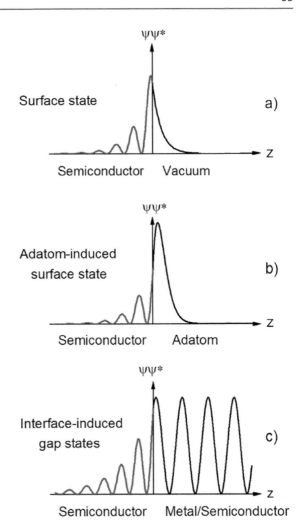

or

$$(\hbar^2/2m_0)q^2 = -[W(q) - V_0] - W_1 \pm \{V_1^2 + 4W_1[W(q) - V_0]\}^{1/2}. \qquad (3.32)$$

The diagram in the middle of Fig. 3.14 illustrates the variation of the imaginary part q of the complex wave vector (3.28) across the band gap. The complex band structure of a one-dimensional solid thus consists of a loop of *virtual* gap states (Heine, 1965), ViGS for short, with complex wave vectors between the two bulk bands at the *Brillouin*-zone boundary π/a.

The existence of *real* surface and interface states then requires additional boundary conditions. At the edges of the of the bulk energy bands, the imaginary part q of the wave vector vanishes, and it passes through a maximum of

$$(\hbar^2/2m_0)q_{max}^2 = V_1^2/4W_1 \tag{3.33}$$

at $V_1^2/4W_1$ below mid-gap. The reciprocal $1/q$ is the decay length of the respective ViGS wave-functions into the linear chain. The maximum value q_{max} just below the center gap position gives the minimum decay length. Towards the edges of the bulk bands, $1/q$ approaches infinity, i.e., the wave-functions become delocalized and behave like *Bloch waves*.

For a linear chain, the density of states of the virtual gap states results as (Garcia-Moliner & Flores, 1979)

$$D_{gs}^{vi}(W) = 1/2\pi[V_1^2 - (W - V_0 - W_1)^2]^{1/2}dW. \tag{3.34}$$

It varies U-shaped across the band gap, as shown in the rightmost diagram in Fig. 3.14.

It should be explicitly noted that the variations of the decay lengths (3.33) and the density of states (3.34) of the gap states with complex wave vectors (3.28) are, first of all, *mathematical* solutions of Schrödinger's equation for a one-dimensional solid with a simple cosine potential. This is why Heine (1965) had called them *virtual* gap states.

The existence of *real* clean-surface and adatom-induced surface states as well as of interface-induced gap states requires additional boundary conditions. Maue (1935) had already shown in his fundamental article that a real surface state will exist at the end of the chain only if the potential $V(z)$ is attractive at the matching position of the respective wave-function $\psi_q(z)$ and its exponential tail into vacuum (see Fig. 3.15a). The predominant character of the gap states changes across the gap from donor-like in the lower half to acceptor-like in the upper half of the gap as the valence band and the conduction band, respectively, provide the induced gap states (Tejedor et al., 1977). Hence, Tejedor et al. inferred that the energy level $W(q_{max})$ of the minimum decay length $1/q_{max}$ of the virtual and then also of the induced gap states at $V_1^2/4W_1$ below mid-gap is the charge neutrality level or, as it is also called with regard to the changing character of the gap states, branch point W_{bp} of the induced gap states. At W_{bp}, the contributions from the valence and conduction bands to the induced gap states are of equal magnitude.

3.7   Metal–Semiconductor Contacts and MIGS

Applying Heine's (1965) concept Tejedor et al. (1977) identified the continuum of interface states introduced by Cowley and Sze (1965) by a continuum of metal- or, more generally, interface-induced gap states. Their detailed analysis revealed the barrier-height of *n*-type *Schottky* contacts to be (see also Flores et al., 1987)

$$\Phi_{Bn}^{T-F} = W_g - \Phi_{bp}^p - S_\Phi(\chi + W_g - \Phi_{bn}^p - \Phi_m), \tag{3.35}$$

where $\Phi_{bp}^{p} = W_{bp} - W_{v}$ is the p-type branch-point energy of the continuum of the interface-induced gap states. The p-type and the n-type branch-point energy, $\Phi_{bp}^{n} = W_{c} - W_{bp}$, add to the width of the band gap W_{g} so that Eq. (3.35) can be rewritten:

$$\Phi_{Bn}^{T-F} = \Phi_{bp}^{n} - S_{\Phi}\left(I - \Phi_{bn}^{P} - \Phi_{m}\right)$$

$$= \Phi_{bp}^{n} - S_{\Phi}\left(W_{vac} - W_{bp} - \Phi_{m}\right)$$

$$= \Phi_{bp}^{n} + S_{\Phi}\left(\Phi_{m} - \Phi_{bp}\right). \tag{3.36}$$

Obviously, the relations (3.7) and (3.36) are identical. The difference, however, is that relation (3.36) is now based on a *physical* concept and no longer just on a *phenomenological* model.

Like many others, Tejedor et al. (1977) overlooked the fact that Heine (1965) had already obtained relation (3.36) in his pioneering paper,

$$\Phi_{Bn} = \Phi_{CNL}^{n} + S_{\Phi}(\varphi_{m} - \varphi_{s}), \tag{3.37}$$

but with an important difference. Heine emphasized that in the dipole $S_{\Phi}(\varphi_{m} - \varphi_{s})$ term the quantities φ_{m} and φ_{s} are "*not quite the real work functions, but the volume contributions to them without the surface dipoles on the free surface*". Based on the discussions in Sect. 3.5.2 and 3.5.3, the dipole term $S_{\Phi}(\Phi_{m} - \Phi_{bp})$ in Eqs. (3.36) and (3.37) is therefore replaced by $S_{X}(X_{m} - X_{semi})$ in the following.

Cowley and Sze (1965) had postulated an interfacial layer between the metal and the semiconductor that separates the dipole charges on the metal side and in the interface states on the semiconductor side of the contact, thus accommodating the corresponding potential drop. They assumed this interface layer to have a thickness of a few Ångstroms and to be transparent to electrons whose energy is greater than the potential barrier. The concept of metal-induced gap states, however, requires no such interlayer because the metal wave-functions directly tunnel into the semiconductor. In the slope parameter (3.6), the thickness δ_{is} of the separating layer can therefore be well approximated by the $2/q_{max}$.

3.8   Semiconductor Heterostructures

3.8.1   Anderson's Rule

Anderson (1962) explained the band lineup at intimate and abrupt semiconductor heterostructures in terms of the *Schottky–Mott* rule, i.e., no interaction at the interface formation and no interface states were considered. Figure 3.16 schematically shows a corresponding band diagram directly at the interface, i.e., any band bending is neglected.

Fig. 3.16 Schematic band
diagram of a intimate and
abrupt semiconductor
heterostructure for the case of
no interface states

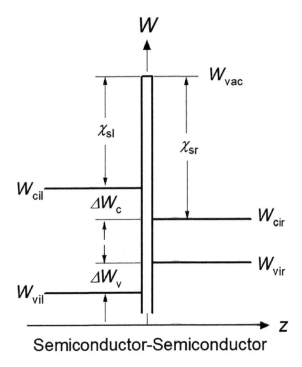

Semiconductor-Semiconductor

He obtained the conduction-band offset

$$\Delta W_c = W_{cil} - W_{cir} = \chi_{sr} - \chi_{sl} \tag{3.38}$$

as the difference of the electron affinities and the valence-band discontinuity

$$\Delta W_v = W_{vir} - W_{vil} = I_l - I_r \tag{3.39}$$

as the difference of the ionization energies $I = W_{vac} - W_{vi}$ of the two semiconductors in contact.

3.8.2 Heterostructures and Interface-Induced Gap States

As the early *Schottky–Mott* rule, Anderson's rule also considers surface rather than bulk properties and no interface states are considered. As early as 1975, Kroemer discussed at length the problems encountered in the theory of heterostructure offsets. He ended up with the conclusion that, as he wrote in a later contribution (Kroemer, 1985), Anderson's rule *"might be a practical 'cookbook recipe' with some utility, from the fundamental point-of-view of 'understanding' the physics of the band lineups, the rule continues to be no answer at all*". He proved that heterostructure discontinuities are bulk rather than surface properties and proposed a *"new bulk materials parameter that in many ways takes over the role of the electron affinity"*. Although he realized quantum-mechanically tunneling in the valence-band offset region, he did not refer to the associated interface-induced gap states that had been already proposed by Heine (1965) for *Schottky* contacts.

It remained for Tejedor and Flores (1977) to apply Heine's (1965) concept of interface-induced gap states to semiconductor heterostructures as well. Right from the beginning of their study they stated that they *"are interested in relating ΔW_v to both semiconductor electron affinities"*. The result of their detailed study is that the valence-band discontinuity can be written as (see also Flores et al., 1987)

$$\Delta W_v = \Phi_{bpl}^p - \Phi_{bpr}^p + S_\Phi\left(\chi_l - W_{gl} - \Phi_{bpl}^p - \chi_r - W_{gr} + \Phi_{bpr}^p\right). \tag{3.40}$$

Considering that the ionization energy and the electron affinity differ by the width of the band gap, i.e., $I = \chi + W_g$, and the definition of the CNL or, now, branch-point work-function $\Phi_{bp} = W_{vac} - W_{bp}$, Eq. (3.40) can be rewritten:

$$\Delta W_v = \Phi_{bpl}^p - \Phi_{bpr}^p + S_\Phi\left(I_l - \Phi_{bpl}^p - I_r + \Phi_{bpr}^p\right)$$
$$= \Phi_{bpl}^p - \Phi_{bpr}^p + S_\Phi\left(\Phi_{bpl} - \Phi_{bpr}\right). \tag{3.41}$$

Like the term $S_\Phi\left(\Phi_m - \Phi_{bp}\right)$ in Eqs. (3.7) and (3.36), the term $S_\Phi\left(\Phi_{bpl} - \Phi_{bpr}\right)$ in Eq. (3.41) also represents the interface-induced dipole. The discussions of the adatom-induced core level shifts on semiconductor surfaces in Sect. 3.5.2 led to the conclusion that the difference $\Phi_m - \Phi_{CNL}$ of the metal work- function and the CNL work-function of the interface states should be replaced by the difference $X_{sl} - X_{sr}$ of the metal and the semiconductor electronegativities. Consequently, the valence band offsets are finally given by:

$$\Delta W_v = \Phi_{bpl}^p - \Phi_{bpr}^p + S_X(X_{sl} - X_{sr}). \tag{3.42}$$

The barrier heights of *Schottky* contacts are determined by the branch-point energies Φ_{bp}^p and/or Φ_{bp}^n and the slope parameter $S_X = d\Phi_{Bn,p}^{hom}/dX_{Mied}$ of the respective semiconductor. Both quantities can thus be easily obtained from plots of experimental barrier heights Φ_{Bn}^{hom} or Φ_{Bp}^{hom} versus the differences $X_m - X_s$ of the metal and semiconductor electronegativities without any further assumptions.

In comparison to *Schottky* contacts, the analysis of the band offsets of semiconductor heterostructures is somewhat more complicated because not two but three quantities are involved, namely the branch-point energies Φ_{bpl}^p and Φ_{bpr}^p of the two semiconductors and the slope parameter S_X of the semiconductor with the larger branch-point energy. However, the simple rearrangement of terms

$$\Delta W_v - S_X(X_{sl} - X_{sr}) = \Phi_{bpl}^p - \Phi_{bpr}^p \tag{3.43}$$

provides a way out of the dilemma. For different heterostructures on the same substrate semiconductor, the difference $\Delta W_v - S_X(X_{sl} - X_{sr})$ will vary linearly as a function of the p-type branch-point energies of the other semiconductors. The linear fit to the data should then have a slope of 1, and the intercept with the ordinate is the branch-point energy of the substrate semiconductor.

3.8.3 Determination of Valence-Band Offsets

Valence-band offsets are most reliably, and therefore most commonly, determined from the energies of core-level lines in X-ray photoemission spectra (XPS) taken with bulk samples of the two semiconductors forming the heterostructure and with the respective interface itself. This technique was developed and first described by Kraut et al. (1980). The valence-band discontinuity results as

$$\Delta W_v = W_{vir} - W_{vil}$$

$$= W_{ir}(nl) - W_{il}(\tilde{n}\tilde{l}) + [W_{vbr} - W_{br}(nl)] - [W_{vbl} - W_{bl}(\tilde{n}\tilde{l})], \qquad (3.44)$$

where (nl) and $(\tilde{n}\tilde{l})$ denote the core levels considered with the semiconductors on the left, l, and the right, r, side of the interface, respectively. The subscripts i and b again characterize interface and bulk properties, respectively. The escape depths of the respective photoelectrons are only of the order of 2 nm and therefore the energy difference $W_{ir}(nl) - W_{il}(\tilde{n}\tilde{l})$ between the core levels at the interface is determined from the X-ray photoemission spectra recorded during respective interruptions during the growth of the heterostructure. The energy differences between the core levels in the bulk, $W_{vbr} - W_{br}(nl)$ and $W_{vbl} - W_{bl}(\tilde{n}\tilde{l})$, of the two semiconductors are evaluated separately.

Interface-Induced Gap States

4

4.1   Calculated Branch-Point Energies

Flores and his co-workers (Louis et al., 1976; Tejedor et al., 1977; Tejedor & Flores, 1978) were the first who estimated the branch-point energies of Si (0 eV), Ge (0.17 eV), AlSb (0.77 eV), GaP (0.77 eV), GaAs (0.55 eV), GaSb (0.61 eV), and InAs (0.45 eV). Much like in Maue's (1935) one-dimensional model, they calculated the branch-point energies as the average of the mid-gap energies at the X- and L-points of the *Brillouin* zone for face-centered cubic crystals. They applied the band structures calculated using the empirical pseudopotential method (EPP).

In a series of articles, Tersoff (1984a, 1985, 1986) then reapplied Heine's (1965) concept of interface-induced gap states to explain the barrier heights of *Schottky* contacts and the band offsets of heterostructures. However, he did not refer to the preceding and even more extensive work of Tejedor and Flores (1978). To Tersoff's credit, however, he was the first to calculate the branch-point energies of 16 semiconductors, which were in the focus of interest at that time. For his calculations, he applied the linearized augmented plane-wave method (APW). Principally, Tersoff followed the same line as *Flores* and his co-workers. But he summed over 152 points in the 1st Brillouin zone, instead of two like Flores and co-workers. Since the local density approximation (LDA) used predicts band gaps 30 to 50% smaller than the experimental values, Tersoff applied the "*scissors operator*" (Baraff & Schlüter, 1984; Godby et al., 1988), which compensates for the differences by a rigid upward shift of the conduction bands. The *p*-type branch-point energies obtained by *Tersoff* are compiled in Table 4.1.

Tersoff's branch-point energies turned out to be a reliable set of data as will be shown in the sections to follow. However, they require a lot of computational effort. Cardona and Christensen (1987) were the first to apply Baldereschi's (1972, 1973) concept of mean *k*-points in the *Brillouin* zone. This approach appears plausible since the barrier heights

© The Author(s), under exclusive license to Springer Nature Switzerland AG 2024   61
W. Mönch, *Electronic Structure of Semiconductor Interfaces*, Synthesis Lectures on Engineering, Science, and Technology, https://doi.org/10.1007/978-3-031-59064-1_4

Table 4.1 Calculated branch-point energies $\Phi_{bp}^p = W_{bp} - W_v$ in eV

	Tersoff	Mönch	Brudnyi	Robertson Falabretti I Guo		Bechstedt	Hinuma	Ghosh	Varley
C		1.77	2.22				1.85		1.45
Si	0.36	0.03	0.37	0.30	0.20	0.29	0.16	0.25	0.14
Ge	0.18	−0.28	0.06		0.10		−0.28	−0.21	−0.23
3C-SiC		1.44	1.76	1.61					1.66
4H-SiC			1.76						
6H-SiC			1.76	1.90					
3C-BN		3.25	3.73						3.52
3C-AlN		2.97				3.42			3.10
2H-AlN			3.48	3.97	3.08	3.35			
AlP	1.27	1.13	1.31		1.30		1.24	1.20	1.38
AlAs	1.05	0.92	1.07	0.92	0.83		0.94	0.89	0.96
AlSb	0.53	0.53	0.45		0.53		0.35	0.33	0.40
3C-GaN		2.37				2.37	2.45		2.23
2H-GaN			2.66	2.88	2.14	2.43			
GaP	0.81	0.83	1.03	0.83	0.83	0.74	0.65	0.78	0.84
GaAs	0.50	0.52	0.70	0.50	0.50	0.54	0.35	0.43	0.43
GaSb	0.07	0.16	0			0.16	−0.18	−0.07	−0.08
3C-InN		1.51				1.49			1.35
Q2			1.76	1.87		1.58			
InP	0.76	0.86	0.90		0.93	0.82	0.72	0.83	0.77
InAs	0.50	0.50	0.50	0.69	0.45	0.59	0.40	0.50	0.42
InSb	0.01	0.22	0.05	0.22		0.28	−0.03	0.07	0.04
ZnO				3.27	3.27	3.40	3.54		
ZnS		2.05			1.75		2.13		2.04
ZnSe	1.70	1.48			1.60		1.66	1.57	1.52
ZnTe	0.84	1.00	1.72		1.09		0.89	0.83	0.87
CdO						2.45			
CdS		1.93			1.90		1.94		1.73
CdSe		1.53			1.50		1.55	1.55	1.40
CdTe	0.85	1.12	1.56		1.12		0.98	0.93	0.91
GaS			1.02						

(continued)

Table 4.1 (continued)

	Tersoff	Mönch	Brudnyi	Robertson Falabretti I Guo	Bechstedt	Hinuma	Ghosh	Varley
GaSe			0.80					
GaTe			0.72					
InSe			0.77					
CuGaS$_2$		1.43						
CuGaSe$_2$		1.25						
CuGaTe$_2$		0.61						
Al$_2$O$_3$				6.00	5.50			
Ga$_2$0$_3$				2.80				
In$_2$O$_3$				3.30	3.30	3.79		
SiO$_2$				4.50	4.50	4.52		
SrTiO$_3$				2.30				

of metal–semiconductor contacts do not depend on the crystal orientation and the crystal structure of the semiconductor substrate if epitaxial contacts are not considered (see Figs. 2.19a and 2.23, respectively). Cardona and Christensen also applied the LD approximation in their calculations but avoided the use of the scissors operator for adjusting the too-small band gaps. Instead, they increased the band gap at the mean-value k-point by taking the energy distance between the average of the top two valence bands and the bottom two conduction bands. Because of this rather arbitrary procedure, their numerical results are not considered here.

Mönch (1996) then combined Baldereschi's concept (1972, 1973) of mean-value k-points and Penn's (1962) model of isotropic semiconductors by equating the energy gap at the mean-value k-point with Penn's average or dielectric band gap W_{dg}. For face-centered cubic lattices the best mean-value point is $k_{mv} = (0.6223,02,953,0){\cdot}(2\pi/a)$ (Baldereschi, 1972, 1973; Chadi and Cohen, 1973). The generalization of the one-dimensional ViGS model outlined in Sect. 3.6 to three dimensions suggests, that the branch point of the ViGS continuum is found slightly below the middle of the dielectric gap W_{dg} at the mean-value point k_{mv}. As illustrated in Fig. 4.1, one thus expects $W_{bp} \approx W_v(k_{mv}) + W_{dg}/2$ and

$$\Phi_{bp}^p = W_{bp} - W_v(\Gamma) \approx W_{dg}/2 - [W_v(\Gamma) - W_v(k_{mv})], \qquad (4.1)$$

where $W_v(\Gamma)$ is the valence-band maximum in the middle of the 1st *Brillouin* zone.

Penn (1962) described semiconductors by an isotropic three-dimensional nearly-free electron model. He got the optical dielectric function, i.e., the electronic contribution to the static dielectric constant,

Fig. 4.1 Dispersion of the top valence (vb) and the lowest conduction band (cb) around the mean-value $\underline{k}$-point $\underline{k}_{mv}$ (schematically). The width of the energy gap at the mean-value point equals the dielectric band gap W_{dg}. From Mönch (2004)

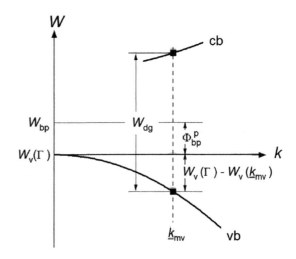

$$\varepsilon_\infty - 1 = (\hbar\omega_p/W_{dg})^2\left[1 - W_{dg}/4W_1 + (W_{dg}/4W_1)^2/3 - \ldots\right], \qquad (4.2)$$

where W_1 was already defined in Eq. (3.27), and

$$\hbar\omega_p = (e_0^2 n_b/\varepsilon_0 m_o)^{1/2} \qquad (4.3)$$

is the plasmon energy of the bulk electrons of density n_b. For the group-IV elemental and the III–V and zinc-chalcogenide compound semiconductors, $W_{dg}/4W_1$ varies between 0.07 and 0.15, as already Penn had noted. Thus, the second and third term in the square brackets of (4.2) can be neglected, yielding the dielectric gap

$$W_{dg} = \hbar\omega_p \big/ (\varepsilon_\infty - 1)^{1/2} \qquad (4.4)$$

Mönch (1996) calculated the dispersion $W_v(\Gamma) - W_v(k_{mv})$ of the top valence band using the empirical tight-binding approximation (ETB) and used experimental values of the bulk plasmon energy $\hbar\omega_p$ and the optical dielectric constant ε_∞ to determine the dielectric band gap W_{dg}.

Meanwhile, quasiparticle (QP) band structure calculations based on the GW approximation of the self-energy operator have become the "state of the art" in most advanced electronic structure calculations for semiconductors (Rohlfing et al., 1995) and have solved the LDA band gap dilemma. The QP-GW band gaps deviate by in the order of less than 0.2 eV or 7.9% from the experimental values (Bechstedt et al., 2009). Therefore, ETB valence band dispersions $W_v(\Gamma) - W_v(k_{mv})$ and the dielectric band gaps used by Mönch (1996) as input parameters are compared with the respective QP-GW results in Figs. 4.2 and 4.3, respectively. The valence band energies and the band gaps at the mean-value $\underline{k}$-point are not of general interest, but were calculated specifically for diamond, Si, Ge, SiC, GaAs, CdS, and MgO by Rohlfing et al. (1993, 1995) and Baumeier

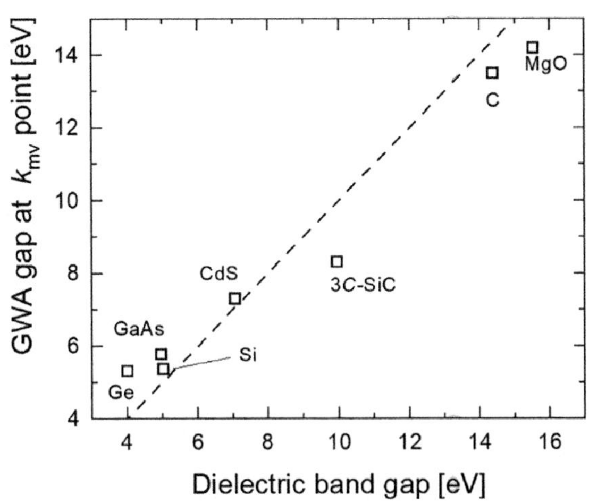

Fig. 4.2 Width of the band gap at the mean-value point of diamond, Si, Ge, SiC, GaAs, and CdS as calculated in the GW approximation by Rohlfing et al. (1993, 1995) and Baumeier et al. (2007) plotted versus the width of dielectric band gap

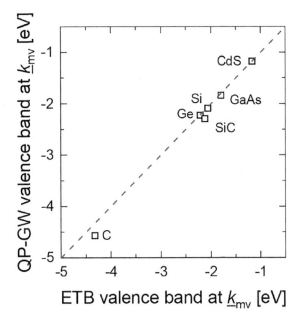

Fig. 4.3 Position of the top valence band at the mean-value point relative to the valence-band maximum: Comparison of OP-GW data from Rohlfing et al. (1993, 1995) and ETB data from Mönch (1996). The dashed line is a least-squares fit and has a slope parameter of 1.08 ± 0.03

et al. (2007), respectively. The dashed lines in Figs. 4.2 and 4.3 demonstrate good agreement between the respective data sets, so the ETB and experimental data can be used here without any limitations.

As a further check, the branch-point energies $W_{bp} - W_v(k_{mv}) = [W_v(\Gamma) - W_v(k_{mv})]_{\text{ETB}} + [\Phi_{bp}^p]_{\text{APW}}$ are plotted versus the widths of the dielectric band gaps W_{dg} in Fig. 4.4, where $[\Phi_{bp}^p]_{\text{APW}}$ are the branch-point energies calculated by Tersoff (1986).

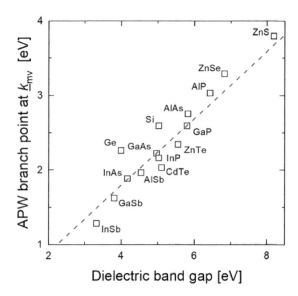

Fig. 4.4 Position of the APW branch-point above the valence band at the mean-value point versus the dielectric band gap. The dashed line is a least-squares fit and has a slope parameter of 0.449 ± 0.007

The dashed line is the linear least-squares fit

$$W_{bp} - W_v(k_{mv}) = (0.449 \pm 0.007) \cdot W_{dg}. \tag{4.5}$$

As predicted by the simple one-dimensional model discussed by Maue (1935), the branch-points of the virtual gap states are indeed slightly below the middle of the dielectric gap by 5%. Only the data points of the elemental semiconductors Si and Ge deviate somewhat from the general trend. On this basis, Mönch (1996) then calculated the branch-point energies of the group IV element semiconductors, SiC, and the III–V, II–VI and, later on (2006), some of the I–III–VI$_2$ compound semiconductors with the relation

$$\Phi_{bp}^p = W_{bp} - W_v(\Gamma) = 0.449 \cdot W_{dg} - [W(\Gamma)_v - W_v(k_{mv})] \tag{4.6}$$

and the ETB valence-band dispersions $[W_v(\Gamma) - W_v(k_{mv})]_{\text{ETB}}$. The second column of Table 4.1 shows Mönch's calculated branch-point energies.

A large step forward was made by Bechstedt and co-workers (Schleife et al., 2009; Belabbes et al., 2011, 2012; Höffling et al., 2012). They computed QP-GW band structures and obtained the branch-point energies as the Brillouin zone averages of the mid-gap energies between the topmost valence and the lowest conduction bands. Their data for Si, 12 of the III–V and II–VI compounds, SiO$_2$, In$_2$O$_3$, and MgO are compiled in the sixth column of Table 4.1. Figure 4.5 compares the branch-point energies computed by Bechstedt and co-workers using the most advanced QP-GW theories and the corresponding values calculated by Mönch who applied the much simpler ETB approach. The dashed

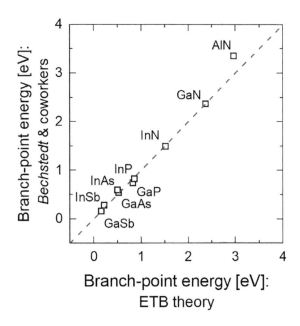

Fig. 4.5 Branch-point energies from QP-GW calculations versus values calculated using ETB theory from Bechstedt and coworkers (Schleife et al., 2009; Belabbes et al., 2011, 2012; Höffling et al., 2010), and Mönch (1996), respectively

line shows that the results of both theoretical approaches agree excellently. Only the values for AlN deviate by 0.3 eV, i.e., by 10%. Because the ETB approach is less involved than the computational demanding QP-GW theory, ETB branch-point energies were calculated for SiC, almost all III–V and II–VI as well as some I–III–VI$_2$ compounds (Mönch, 1996, 2006). Therefore, these ETB branch-point energies will be used later in the analysis of experimental valence-band offsets in Chap. 5.

Hinuma et al. (2014) also calculated quasi-particle band structures and the branch-point energies of diamond, Si, Ge, and 11 of the III–V and 7 of the II–VI compounds. They used the GWΓ^1 approximation that includes specific corrections in the self-energy. The branch-point energies obtained are compiled in Table 4.1 and plotted in Fig. 4.6 versus the corresponding ETB data of Mönch (1996). The linear least-squares fit has a slope of 1.14 ± 0.04, i.e., the GWΓ^1 data are about 14% lager than the ETB values. This difference can be attributed, at least in part, to the assumption of Hinuma et al. that the branch points are in the middle of the of the average gaps rather than a few percent below them.

Brudnyi et al. (1995, 2008, 2015, 2017) presented a further, large set of branch-point energies. They calculated the band structures of the semiconductors considered using an empirical pseudopotential method (EPP) and then obtained the branch points as the center of the average gaps between the lowest conduction and the highest valence bands at 10 special k-points in the 1st *Brillouin* zone. They specifically pointed out that the average gaps are close to Penn's dielectric band gaps (4.4). Their data are also compiled in Table 4.1. In Fig. 4.7 these data are plotted versus Mönch's (1996) ETB branch-point energies. The linear least-squares has a slope of 1.19 ± 0.06, i.e., the EPP data are about

Fig. 4.6 QP-GWΓ[1]
branch-point energies
calculated by Hinuma et al.
(2014) plotted versus
corresponding ETB values
calculated by Mönch (1996).
The dashed line is the linear
least-squares fit
$$\Phi^P_{bp} = -(0.26 \pm 0.054) +$$
$$(1.17 \pm 0.04)\Phi^P_{bp}(\text{ETB})[\text{eV}]$$
to the data

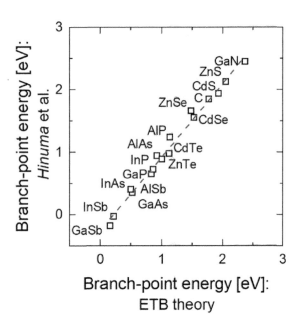

Fig. 4.7 EPP branch-point
energies reported by Brudnyi
et al. (1995, 2008, 2017)
plotted versus corresponding
ETB values calculated by
Mönch (1996). The dashed line
is the linear least-squares fit to
the data with a slope of
1.19 ± 0.05

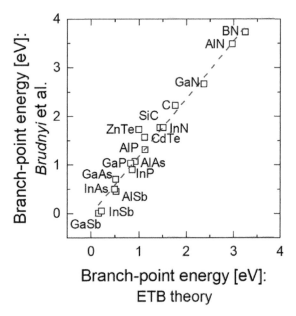

20% lager than the ETB values. This difference can also be attributed, at least in part, to the assumption that the branch points are in the middle of the of the average gaps rather than a few percent below them.

Fig. 4.8 PP branch-point energies reported by Robertson and Falabretti (2006) plotted versus corresponding ETB values calculated by Mönch (1996). The dashed line is the linear least-squares fit $\Phi_{bp}^P = (1.26 \pm 0.04)\Phi_{bp}^P(\text{ETB})[\text{eV}]$ to the data

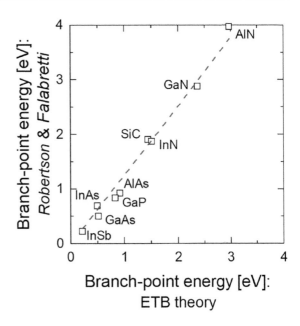

Robertson and co-workers (Robertson and Chen, 1999; Peakock and Robertson, 2002; Robertson and Falabretti, 2006) used the same methods as Tersoff (1984) and calculated the branch-point energies of a great variety of semiconductors as, for example, Si, Ge, SiC, the III–V compounds, and many oxides, silicates, and titanites. They used band structures obtained by the pseudopotential method (PP) and again adjusted the band gaps to the experimental values using the scissors operator. Most of their results are also compiled in Table 4.1. Some of these data (Robertson and Falabretti, 2006) are plotted in Fig. 4.8 versus Mönch's (1996) corresponding ETB branch-point energies. The linear least-squares fit has a slope of 1.26 ± 0.04, i.e., the EPP data are almost 30% lager than the ETB values.

To avoid the scissors operator, Guo et al. (2019) used screened exchange hybrid functionals (SX). For Si, Ge, and most of the III–V and II–VI compound semiconductors, the SX band gaps agree well with the experimental values. The SX branch-point energies are listed in Table 4.1 and plotted versus Mönch's (1996) corresponding ETB branch-point energies in Fig. 4.9. Obviously, the SX data excellently agree with Mönch's data.

Mourad (2012, 2013) computed the branch-point energies of AlN, GaN, InN, ZnTe, and CdS. He used an ETB model and calculated the branch-points as an average over the 1st *Brillouin* zone of the mid-gap energy. The data are again in close agreement with the branch-point energies of Mönch (1996). Only the difference of the AlN data is with 0.3 eV somewhat larger.

Ghosh et al. (2022) bypassed computationally intensive QP-GW calculations to avoid underestimation of bulk band gaps by investigating the performance of several advanced exchange–correlation functionals. They found the Local Modified Becke-Johnson (LMBJ)

Fig. 4.9 SX branch-point
energies reported by Guo et al.
(2019) plotted versus
corresponding ETB values
calculated by Mönch (1996).
The dashed line is the linear
least-squares fit $\Phi^P_{bp} =$
$(1.00 \pm 0.02)\Phi^P_{bp}(ETB)[eV]$
to the data

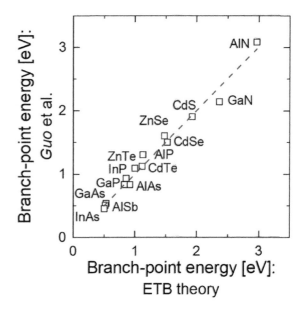

Fig. 4.9 SX branch-point
energies reported by Guo et al.
(2019) plotted versus
corresponding ETB values
calculated by Mönch (1996).
The dashed line is the linear
least-squares fit $\Phi^P_{bp} =$
$(1.00 \pm 0.02)\Phi^P_{bp}(ETB)[eV]$
to the data

functional to be the best one. Their calculated branch-point energies are compiled in
Table 4.1 and plotted versus Mönch's (1996) corresponding ETB branch-point energies
in Fig. 4.10. The linear lest-squares fit reveals good agreement between the data calculated
by Ghosh et al. and Mönch's (1996) ETB branch-point energies.

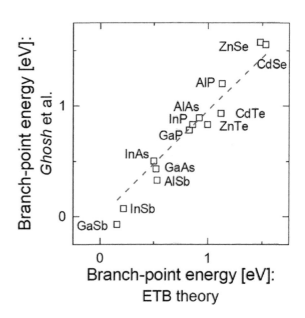

Fig. 4.10 LMBJ branch-point
energies reported by Ghosh
et al. (2022) plotted versus
corresponding ETB values
calculated by Mönch (1996).
The dashed line is the linear
least-squares fit $\Phi^P_{bp} =$
$(0.96 \pm 0.04)\Phi^P_{bp}(ETB)[eV]$
to the data

Fig. 4.11 VASP branch-point energies calculated by Varley et al. (2024) plotted versus corresponding ETB values calculated by Mönch (1996). The dashed line is the linear least-squares fit $\Phi_{bp}^{p} = (0.99 \pm 0.02)\Phi_{bp}^{p}(\text{ETB})[\text{eV}]$ to the data

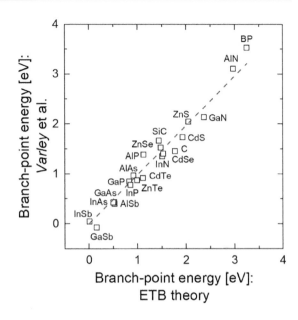

Varley et al. (2024) applied the widely used "Vienna *ab-initio* simulation package" (VASP) to calculate the branch-point energies of the group IV elemental semiconductors as well as of the IV–IV, III–V, and some II–VI compound semiconductors. The last column of Table 4.1 shows their results, with the SiC data corrected.[1] Figure 4.11 compares the *ab-initio* computed data of Varley et al. with the corresponding ETB values of Mönch (1996). The linear least-squares fit shows that these *ab-initio* calculated data confirm Mönch's ETB branch-point energies in an excellent way.

4.2 Slope Parameters

Cowley and Sze (1965) basically modelled the interface of a metal–semiconductor contact by a dipole layer of thickness δ_{is} which separates the surface charges in a continuum of interface states of the semiconductor and on the metal. They derived the slope parameter as (see. Eq. 3.6)

$$S_\Phi = \mathrm{d}\Phi_{Bn}/\mathrm{d}\Phi_m = 1/(1 + e_0^2 \delta_{is} D_{is}/\varepsilon_0 \varepsilon_i),$$

---

[1] Private communication by Dr. *J.B. Varley* 06.03.2024.

where δ_{is} and D_{is} are the thickness of the dipole layer and the density of states of the continuum of interface states, respectively. The concept of metal- or, to include also heterostructures, interface-induced gap states, however, requires no such charge-separating interlayer because within the semiconductor band gap the metal wave-functions directly tunnel into the semiconductor. In the slope parameter, the thickness δ_{is} of the charge-separating interlayer can then be well approximated by the by the decay length $1/2q_{gs}^{mi}$ of the interface states at their branch point W_{bp}. Furthermore, the difference of the metal work-function and the CNL work-function of the interface states was replaced by the difference $X_m - X_s$ of the metal and semiconductor electronegativities; see Sect. 3.5.2. Consequently, the slope parameter is obtained as

$$S_X = d\Phi_{Bn}/dX_m = A_X/[1 + e_0^2 D_{gs}^{mi}(W_{bp})/2\varepsilon_0\varepsilon_i q_{gs}^{mi}(W_{bp})], \qquad (4.7)$$

where $A_X = 0.86$ eV/Miedema unit when Miedema's electronegativities are used.

Calculated decay lengths $1/2q_{gs}^{mi}(W_{bp})$ and densities of state $D_{gs}^{mi}(W_{bp})$ of the metal-induced gap states at their branch points are summarized in Table 4.2. Already the simple one-dimensional model discussed by Maue (1935) revealed that both quantities vary proportional to the width of the band gap; see Eqs. (3.33) and (3.34). The combination of Baldereschi's (1972, 1973) mean-value k-point concept and Penn's (1962) model of isotropic semiconductors led to a set of theoretical branch-point energies (Mönch, 1996) that were later confirmed by calculations of Bechstedt and co-workers (Schleife et al., 2009; Belabbes et al., 2011, 2012; Höffling et al., 2010; Hinuma et al., 2014), and Ghosh et al. (2022) who used the most advanced quasiparticle-GW theories. Penn's dielectric band gap (4.4) varies inversely proportional to the square root of the optical susceptibility $\varepsilon_\infty - 1$. Therefore, the theoretical $e_0^2 D_{gs}^{mi}(W_{bp})/2\varepsilon_0 q_{gs}^{mi}(W_{bp})$ data are plotted versus $\varepsilon_\infty - 1$ of the respective semiconductors in Fig. 4.12. The dashed line is the linear least squares fit

$$e_0^2 D_{gs}^{mi}(W_{bp})/2\varepsilon_0 q_{gs}^{mi}(W_{bp}) = 0.27(\varepsilon_\infty - 1)^{1.91\pm0.14} \qquad (4.8)$$

to the data.

The interface dielectric constant ε_I in Eq. (4.7) can be estimated by modeling the interface dipole of a *Schottky* contact as a parallel-plate capacitor with an *effective* thickness

Table 4.2 Dielectric gap W_{dg}, optical dielectric constants ε_∞, charge-density tailing length $1/2q_{gs}^{mi}$, and density of states D_{gs}^{mi} of metal-induced gap states. The dielectric band gaps were calculated using Eq. (3.47)

Semiconductors	W_{dg} [eV]	ε_∞	$1/2q_{gs}^{mi}$ [nm]	$D_{gs}^{mi} \times 10^{14}$ [cm^{-2} eV^{-1}]
C	14.4	5.7	0.137[d]	2.00[d]
Si	5.04	11.9	0.30[e]	4.50[e] 3.96[b]
Ge	4.02	16.2	0.40[e]	
AlAs	5.81	8.16	0.66[b]	
GaN	10.80	5.8	0.186[g]	
GaP	5.81	9.11	0.294[d]	3.12[a]
GaAs	4.97	10.9	0.30[e] 0.28[c] 0.32[f]	5.00[c] 3.72[a] 2.70[f]
InP	5.04	9.61	0.424[b]	3.45[a]
InSb	3.33	12.8		5.05[a]
ZnS	8.12	5.14	0.09[c] 0.16[a]	1.40[c] 1.96[a]
ZnSe	7.06	5.7	0.19[c] 0.20[b]	2.00[c] 2.28[a]
CdTe	5.11	7.21		2.34[a]
SiO$_2$	23	2.25	0.12[i]	0.15[h]
Al$_2$O$_3$		3.4	0.28[h]	
HfO$_2$		4	0.47[h]	
ZrO$_2$		4.8	0.09[g]	
SrTiO$_3$		5.56	0.20[g]	
LaAlO$_3$		4	0.11[j]	

[a] Louis et al. (1976)
[b] Tejedor et al. (1977)
[c] Louie et al. (1977)
[d] Ihm et al. (1978)
[e] Tersoff (1984a, 1986)
[f] Berthod et al. (1996)
[g] Dong et al. (2006a)
[h] Demkov et al. (2005)
[i] Giustino and Pasquarello (2005)

Fig. 4.12 Theoretical "slope parameters" $(e_0^2/\varepsilon_0)(D_{gs}^{mi}/2q_{gs}^{mi})$ as a function of the optical susceptibility $\varepsilon_\infty - 1$. The calculated values of the densities of states D_{gs}^{mi} and of the decay lengths $1/2q_{gs}^{mi}$ of the MIGS at their branch point are compiled in Table 4.2. The dashed line is the linear least-squares fits to the data

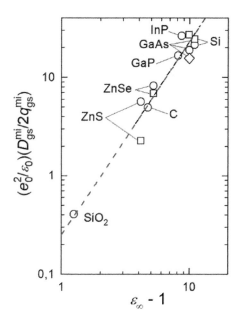

t_{eff} (Heine, 1965; Louie et al., 1977). On the metal and on the semiconductor side of the interface, the charges are distributed within the *Thomas–Fermi* screening length $t_m = L_{\mathrm{Th-F}}$ and the decay length $t_s = 1/2q_{gs}^{mi}$ of the MIGS, respectively. The effective distance is the sum of the true distances divided by the appropriate dielectric screening functions $\varepsilon_m = 1$ and ε_b, respectively, i.e.,

$$t_{eff} = t_s/\varepsilon_i = t_m/\varepsilon_m + t_s/\varepsilon_b. \tag{4.9}$$

A typical value for the *effective Thomas–Fermi* screening length is $t_m/\varepsilon_m = 0.05$ nm, and theoretical decay lengths $1/2q_{gs}^{mi}$ are 0.137 nm for diamond and 0.3 nm for both Si ($\varepsilon_b = 11.9$) and GaAs ($\varepsilon_b = 12.8$) (see Table 4.2). With these values Eq. (4.9) gives interface dielectric constants $\varepsilon_i = 2$ for diamond, 4.6 for Si, and 4 for GaAs. Using $t_m/\varepsilon_m = L_{Th-F} \approx 0.05$ nm and the averages $\bar{t}_s = \overline{1/2q_{gs}^{mi}} \approx 0.25$ nm and $\overline{\varepsilon_b} = 12.5$ of the III–V and II–VI compound semiconductors considered in Table 4.2, relation (4.9) gives $\varepsilon_i \approx 3.5$ as an *estimate* of the mean interface dielectric constant. Substituting this value and the linear least squares fit (4.8) into Eq. (4.7), one then obtains the slope parameter

$$S_X = d\Phi_{Bn}/dX_m \approx A_X/[1 + 0.08(\varepsilon_\infty - 1)^2]$$ (4.10)

of *Schottky* contacts of binary compound semiconductors. As in the case of the branch-point energies, again a bulk property of the semiconductor, its optical dielectric constant ε_∞, determines the other characteristic property of semiconductor interfaces, the slope parameter $S_X = d\Phi_{Bn}/dX_m$.

Comparison of Theoretical and Experimental Data

5

5.1 *p*-Type *Schottky* Contacts

5.1.1 Experimental Data

In Section 3.5.2, the barrier heights of "ideal" *n*- and *p*-type *Schottky* contacts were obtained as

$$\Phi_{Bn}^{hom} = \Phi_{bp}^{n} + S_X(X_m - X_s) \tag{5.1}$$

and

$$\Phi_{Bp}^{hom} = \Phi_{bp}^{p} - S_X(X_m - X_s), \tag{5.2}$$

respectively. With regard to the concept of interface-induced gap states, the CNL-s of the previously unspecified interface states are now identified as branch points of the IFIGS. Thus, the branch-point energies Φ_{bp}^{p} or Φ_{bp}^{n} and the slope parameters S_X can be determined directly, i.e., without any additional assumptions, from experimental barrier heights, provided the interfaces of the metal-semiconductor contacts are "ideal", i.e., abrupt, free of interlayers, clean, and, what has not been explicitly considered for a long time, laterally homogeneous. Unfortunately, the semiconductor surfaces were and still are not well characterized in all cases before the *Schottky* metal is deposited. This lack explains, at least in part, the large scatter of some of the data discussed in the following. The Si *Schottky* contacts considered in Figs. 5.1 and 3.11 were specifically selected because, as the result of very many surface studies, well-documented "HF dips" were used here as the final surface treatment. Details of such procedures can be found, for example, in the books of Mönch (2001, 2004). However, such intense surface studies are not available for all semiconductors.

© The Author(s), under exclusive license to Springer Nature Switzerland AG 2024
W. Mönch, *Electronic Structure of Semiconductor Interfaces*, Synthesis Lectures on Engineering, Science, and Technology, https://doi.org/10.1007/978-3-031-59064-1_5

Fig. 5.1 Barrier heights of laterally homogeneous *p*-diamond and *p*-Si *Schottky* contacts versus the electronegativity difference X_m–X_C and X_m–X_{Si}, respectively. The dashed lines are the linear least-squares fits to the experimental data

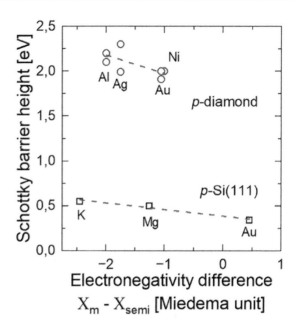

In the following, we first determine experimental Φ_{bp}^{p} values from barrier heights of *p*-type *Schottky* contacts and then compare these with corresponding theoretical *p*-type branch point energies. Experimental barrier heights of *p*-type diamond, Si, SiC, GaN, GaP, GaAs, and III-VI *Schottky* contacts are plotted versus the respective electronegativity differences $X_m - X_s$ in Figs. 5.1, 5.2, 5.3, 5.4, 5.5, 5.6 and 5.7. The dashed lines are the linear least-squares fits to the respective data, and they yield *p*-type branch-point energies Φ_{bp}^{p} and slope parameters S_X of the semiconductors considered. The results of numerous experimental studies have been published for the SiC polytypes and GaN. In selecting the data presented here, we also considered details of the surface preparations prior to the metal deposition that may have favored the formation of extrinsic layers at the interface and that may have then masked the intrinsic interface properties.

Experimental barrier heights of laterally homogeneous *p*-diamond and *p*-Si *Schottky* contacts are plotted versus the electronegativity difference $X_m - X_C$ in Fig. 5.1.[1] The linear least-squares fit to the diamond data

$$\Phi_{Bp}^{\text{hom}} = (1.77 \pm 0.11) - (0.20 \pm 0.07)(X_m - X_C)\,[\text{eV}] \tag{5.3}$$

[1] Figure 5.1. Experimental data of *p*-diamond and *p*-Si *Schottky* contacts from Majdi et al. (2010), Mead and McGill (1976), and Ueda et al. (2013) and Hagemann et al. (2018) and Huba et al. (2009), respectively.

Fig. 5.2 Barrier heights of laterally homogeneous *p*-type 3*C*- (△), 4*H*- (□), and 6*H*-SiC (○) *Schottky* contacts versus the electronegativity difference X_m–X_{SiC}. The dashed line is the linear least-squares fit to the experimental data

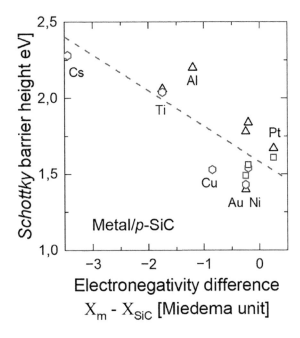

Fig. 5.3 Barrier heights of laterally homogeneous *p*-type GaN *Schottky* contacts versus the electronegativity difference X_m–X_{GaN}. The dashed line is the linear- least squares fit to the experimental data. The Mo data point is not considered in the fit. Data from Rickert et al. (2002) (△), Ueno et al. (2022) (▽), and Aoyama et al. (2022) (□)

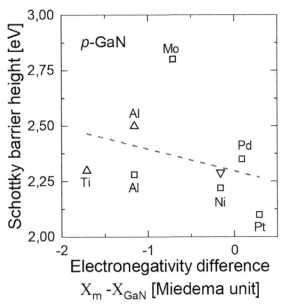

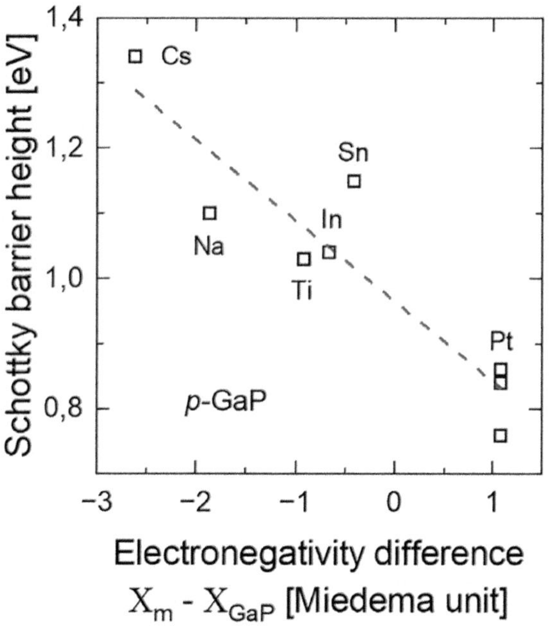

Fig. 5.4 Barrier heights of laterally homogeneous *p*-type GaP *Schottky* contacts versus the electronegativity difference $X_m - X_{GaP}$. The dashed line is the linear- least squares fit to the experimental data. Data from Almeida et al. (1996), v.d.Emde et al. (1982), Evans et al. (1992), and Kusaka et al. (1980)

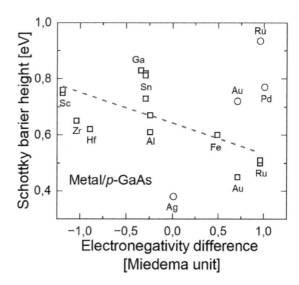

Fig. 5.5 Barrier heights of laterally homogeneous *p*-type GaAs *Schottky* contacts versus the electronegativity difference $X_m - X_{GaAs}$. The dashed line is the linear least-squares fit to the experimental data. The ○ data points are not considered in the fit. Data from Asimov et al. (2013), Cheng et al. (2017), Deniz et al. (2014), Duman et al. (2022), Goodman et al. (1998), Mazari et al. (2002), Myburg et al. (1993, 1998), Sitnitsky et al. (2012), Vereecken & Searson (1999)

Fig. 5.6 Barrier heights of laterally homogeneous *p*-type GaS, GaSe, and GaTe *Schottky* contacts versus the electronegativity difference X_m–X_{Ga-VI}. The dashed lines are the linear- least squares fits to the respective experimental data. Data from Kurtin and Mead (1969), Bose and Pal (1997), Coskun et al. (2003), and Huang et al. (2010)

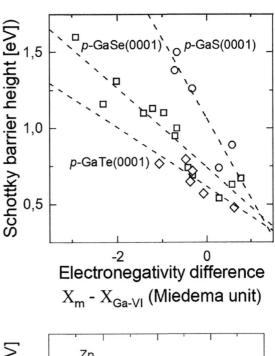

Fig. 5.7 Barrier heights of laterally homogeneous *p*-type InSe *Schottky* contacts versus the electronegativity difference X_m–X_{InSe}. The dashed line is the linear-least squares fit to the experimental data. Data from Brudnyi et al. (2015)

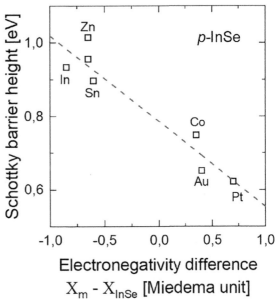

gives the slope parameter $S_X = 0.20$ eV/Miedema unit and the branch-point energy $\Phi_{bp}^p = 1.77$ eV. The latter experimental result agrees exactly with the value calculated by Mönch (1996).

The linear least-squares fit to the p-Si data plotted in Fig.5.1

$$\Phi_{Bp}^{hom} = (0.38 \pm 0.02) - (0.074 \pm 0.014)(X_m - X_{Si})[eV] \tag{5.4}$$

gives the slope parameter $S_X = (0.074 \pm 0.014)$ [eV/Miedema unit] and the branch-point energy $\Phi_{bp}^p = (0.38 \pm 0.02)$ [eV]. The slope parameter agrees with the value obtained from the data of n-Si *Schottky* contacts shown in Fig. 3.11. The n- and p-type branch-point energies $\Phi_{bp}^n = 0.80$ eV and $\Phi_{bp}^p = 0.38$ eV determined from the experimental data shown in Figs. 3.11 and 5.1, respectively, add up to 1.18 eV. Within the margins of experimental error, this value corresponds reasonably well to the bulk band gap of 1.12 eV.

The data of the laterally homogeneous p-type *Schottky* contacts of 3C-, 4H-, and 6H-SiC polytypes plotted in Fig. 5.2[2] show a somewhat larger scatter, but no specific linear dependences of the individual polytypes are apparent. This finding again confirms that within the margins of experimental error these polytypes have identical p-type branch-point energies, as already proposed by Mönch (1994a). Käckell et al. (1994) arrived at the same conclusion. They calculated the band structures of these three SiC polytypes and found that the valence-band offsets of 3C/4H and 3C/6H heterostructures are only 0.05 and 0.02 eV, respectively, i.e., the branch-point energies of the three polytypes are the same. Furthermore, the barrier-heights of laterally homogeneous Pd *Schottky* contacts on n-type 4H- and 6H-SiC measure 1.63 and 1.35 eV, respectively, as shown in Fig. 2.23. The linear least-squares fit to all the data plotted in Fig 5.2

$$\Phi_{Bp}^{hom} = (1.58 \pm 0.06) - (0.23 \pm 0.05)(X_m - X_{SiC})[eV] \tag{5.5}$$

gives the slope parameter $S_X = (0.23 \pm 0.05)$ [eV/Miedema unit] and the branch-point energy $\Phi_{bp}^p = (1.58 \pm 0.06)$ [eV]. The latter result is intermediate between the values of 1.44 eV and 1.76 eV calculated by Mönch (1996) and by Brudnyi and Kosobutsky (2017), respectively.

The experimental barrier heights of p-type GaN, GaP, GaAs, GaS, GaSe, GaTe, and InSe *Schottky* contacts displayed in Figs. 5.3, 5.4, 5.5, 5.6 and 5.7 require no further comments. The ○ GaAs data points in Fig. 5.5 are not considered in the respective fit. Their large deviations from the general trend are most likely due to the preparation of the ohmic back-contacts, which included an annealing treatment immediately prior to the deposition of the *Schottky* metal contact. The linear least-squares fits to the data are

$$\Phi_{Bp}^{hom} = (2.30 \pm 0.10) - (0.10 \pm 0.11)(X_m - X_{GaN})[eV] \tag{5.6}$$

---

[2] Figure 5.2. Experimental data from Aboelfotoh et al. (2003), van Elsbergen et al. (1996), Kojima et al. (2000), Lee et al. (2001), Park et al. (2006), and Satoh & Matsuo (2006).

$$\Phi_{Bp}^{hom} = (0.96 \pm 0.03) - (0.13 \pm 0.02)\,(X_m - X_{GaP})\,[eV] \qquad (5.7)$$

$$\Phi_{Bp}^{hom} = (0.64 \pm 0.03) - (0.11 \pm 0.03)\,(X_m - X_{GaAs})[eV] \qquad (5.8)$$

for the Ga-V compounds and

$$\Phi_{Bp}^{hom} = (1.07 \pm 0.06) - (0.52 \pm 0.12)(X_m - X_{GaS})[eV], \qquad (5.9)$$

$$\Phi_{Bp}^{hom} = (0.74 \pm 0.04) - (0.26 \pm 0.03)(X_m - X_{GaSe})[eV], \qquad (5.10)$$

$$\Phi_{Bp}^{hom} = (0.62 \pm 0.03) - (0.19 \pm 0.05)(X_m - X_{GaTe})[eV], \qquad (5.11)$$

$$\Phi_{Bp}^{hom} = (0.79 \pm 0.02) - (0.23 \pm 0.03)\,(X_m - X_{Inse})[eV] \qquad (5.12)$$

for the III-VI compounds. The experimental branch-point energies are close to or agree with the values calculated by Mönch (1996) for the Ga-V and by Brudnyi et al. (2015) for the III-VI compounds, see Table 4.1.

5.1.2   Branch-Point Energies Φ_{bp}^{p}

In summary, Fig. 5.8 shows the experimental branch-point energies obtained from the barrier heights of *p*-type diamond, Si, SiC, GaN, GaP, GaAs, and four III-VI *Schottky* contacts plotted in Figs. 5.1, 5.2, 5.3, 5.4, 5.5, 5.6 and 5.7 compared with the corresponding theoretical values calculated by Tersoff (1984a) for Si, by Mönch (1996) for diamond, SiC, GaN, GaAs, and GaP, and by Brudnyi et al. (2015) for the III-VI compounds; see Table 4.1. The slope parameter, 0.98 ± 0.05, of the linear least-squares fit indicates excellent agreement between calculated and experimental branch-point energies.

5.1.3   Slope Parameters S_X

In Fig. 5.9, the slope parameters of the laterally homogeneous *p*-type *Schottky* contacts determined from the experimental barrier heights displayed in Figs. 5.1, 5.2, 5.3, 5.4, 5.5, 5.6 and 5.7 for diamond, Si, SiC, GaN, GaP, GaAs, GaS, GaSe, GaTe, and InSe are plotted versus the dielectric susceptibilities $\varepsilon_\infty - 1$ of the semiconductors. The dashed line is the linear least-squares fit

$$A_X/S_X - 1 = (0.27 \pm 0.06)(\varepsilon_\infty - 1)^{1.91 \pm 0.14} \qquad (5.13)$$

Fig. 5.8 Experimental branch-point energies from Figs. 5.1, 5.2, 5.3, 5.4, 5.5, 5.6 and 5.7 versus theoretical branch-point energies calculated for Si by Tersoff (1984), for diamond, SiC, GaN, GaP, and GaAs by Mönch (1996), and for the III-VI compounds by Brudnyi et al. (2015). The dashed line is the linear least-squares fit to the data

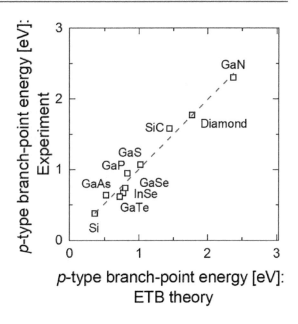

Fig. 5.9 Experimental slope parameters of laterally homogeneous p-type *Schottky* contacts from Figs. 5.1, 5.2, 5.3, 5.4, 5.5, 5.6 and 5.7 versus the dielectric susceptibility $\varepsilon_\infty - 1$. The dashed line is the linear least-squares to the data

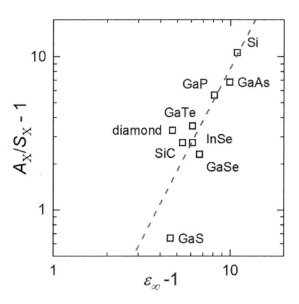

with $A_X = 0.86$ [eV/Miedema unit]. The experimental values confirm the theoretical quadratic dependence on $\varepsilon_\infty - 1$ predicted by Eqs. (4.8) and (4.10), even though the prefactor (0.27 ± 0.6) is somewhat larger than the rough estimate in Eq. (4.10).

Thus, both the branch-point energies Φ_{bp}^p and the slope parameters S_X evaluated from experimental barrier heights of *p*-type diamond, Si, SiC, GaN, GaP, GaAs, GaS, GaSe, GaTe, and InSe *Schottky* contacts agree with the corresponding data calculated under the assumption that metal- or, more generally, interface-induced gap states determine the band-structure lineup at metal semiconductor interfaces. It should be emphasized once again that these values were obtained directly from experimental data without any additional assumptions.

5.2   *n*-Type *Schottky* Contacts and Heterostructures

In the following, we analyze experimental barrier heights of "as ideal as possible" *n*-type *Schottky* contacts and valence-band offsets of heterostructures. As with the *p*-type *Schottky* contacts, discussed in the previous section 5.1, the experimental barrier heights are plotted versus the differences of the metal and semiconductor electronegativities. According to Eq. (5.1), the linear-least squares fits to the data yield the slope parameter S_X and the *n*-type branch-point energy Φ_{bp}^n of the respective semiconductor without any additional assumptions. These values are summarized in Table 5.1.

In Sect. 3.8.2, the valence-band offsets of semiconductor heterostructures were derived as

$$\Delta W_v = \Phi_{bpl}^p - \Phi_{bpr}^p + S_X(X_{sl} - X_{sr}) \tag{5.14}$$

see relation (3.42). Thus, *three* parameters are required to quantitatively describe the valence-band offsets, namely the branch-point energies Φ_{bpl}^p and Φ_{bpr}^p of the two semiconductors involved and the slope parameter S_X of the semiconductor with the larger branch-point energy.

In the following, valence-band discontinuities of different heterostructures with a common semiconductor will be discussed. In the above relation (5.14), the latter semiconductor is chosen as the "right (*r*)" part of the heterostructures. For such heterostructures whose offsets ΔW_v have negative sign, the slope parameter S_{Xr} of the common semiconductor is to be used. For heterostructures with positive ΔW_v values, on the other hand, the different slope parameters S_{Xl} of the respective other semiconductors are relevant.

For plots of the valence-band offsets of various heterostructures with a common semiconductor, the terms in Eq. (5.14) can be conveniently re-arranged as

$$\Delta W_v - S_X(X_{sl} - X_{sr}) = \Phi_{bpl}^p - \Phi_{bpr}^p \tag{5.15}$$

Following Eq. (5.15), the experimental valence-band offsets ΔW_v minus the respective IFIGS electric-dipole contributions $S_X(X_{sl} - X_{sr})$ are plotted versus the branch-point energies Φ_{bpl}^p of the various other semiconductors. Ideally, the linear least-squares fit to the data

Table 5.1 Band gap W_g in eV, optical dielectric constant ε_∞, experimental slope parameters S_X in eV/Miedema unit and branch-point energies Φ_{bp}^n and Φ_{bp}^p in eV obtained from the barrier heights and valence-band offsets of the semiconductors considered in Figs. 4.1 to 4.4 and 4.10 to 4.33

	W_g	ε_∞	Schottky contacts				Heterostructures	
			S_X	Φ_{bp}^p	S_X	Φ_{bp}^n	φ_{vbo}	Φ_{bp}^p
Diamond	5.47	5.7	0.20	1.77	–	–	0.90	1.81
Si	1.12	11.9	0.07	0.38	0.098	0.803	0.99	0.29
Ge	0.66	16.2	–	–	0.050	0.61	0.86	0.05
3C-SiC	2.36	6.38	0.23	1.58	0.13	0.92	1.07	1.45
4H-SiC	3.23	6.38			0.19	1.63		
6H-SiC	3.00	6.38			0.17	1.35		
2H-AlN	6.20	4.84			0.31	2.18	0.93	3.10
2H-GaN	3.39	5.35	0.10	2.30	0.26	1.10	0.91	2.32
GaP	2.27	9.11	0.13	0.96	–	–	1.14	0.83
GaAs	1.42	0.04	0.11	0.64	0.07	0.90	1.02	0.65
GaSb	0.72	14.44			0.034	0.54	0.90	0.06
2H-InN	0.70	5.22	–	–	–	–	1.02	1.54
InP	1.34	9.61	0.10	0.88	0.12	0.48	0.85	0.80
GaS	2.53	5.30	0.52	1.07	–	–	–	–
GaSe	2.01	6.55	0.26	0.74	–	–	–	–
GaTe	1.69	6.2	0.19	0.62	–	–	–	–
InSe	1.26		0.23	0.79	–	–	–	–
Al$_2$O$_3$		3.05	–	–	0.78	4.50	0.97	2.85
Ga$_2$O$_3$	4.85	3.61	–	–	0.34	1.47	--	3.35
SiO$_2$		2.25	–	–	0.75	4.9	1.08	3.33
Si$_3$N$_4$	5.4	3.8	–	–	–	–	1.13	1.64

$$\Delta W_v - S_X(X_{sl} - X_{sr}) = \varphi_{vbo}(\Phi_{bpl}^p - \Phi_{bpr}^p) \tag{5.16}$$

should have a slope $\varphi_{vbo} = 1$ and the ordinate intercept will give the branch-point energy Φ_{bpr}^p of the common semiconductor.

Another rearrangement of terms in Eq. (5.14) appears to be suitable for heterostructures where the common semiconductor has such a large p-type branch-point energy so that all valence-band discontinuities have a negative sign. The transcription

$$\Delta W_v - \Phi_{bpl}^p = -\Phi_{bpr}^p + S_{Xr}(X_{sl} - X_{sr}) \tag{5.17}$$

of Eq. (5.14) then suggests plotting the experimental values of ΔW_v minus the branch-point energies Φ_{bpl}^p of the other semiconductors versus the difference $X_{sl} - X_{sr}$ of the electronegativities of the two semiconductors. The slope parameter and the ordinate intersection of the linear least-squares fits are then equal to the slope parameter S_{Xr} and branch-point energy Φ_{bpr}^p of the common semiconductor, respectively.

In the following analysis of the valence-band offsets of semiconductor heterostructures, we will use the *p*-type branch-point energies of Si and Ge calculated by Tersoff (1984a), of diamond, 3*C*-SiC, the III-V and II-VI compounds calculated by Mönch (1996), and of some of the III-VI compounds calculated by Brudnyi et al. (2015). For all other semiconductors, and insulators are naturally included, branch-point energies evaluated from experimental valence-band offsets will be considered. The slope parameters used are all obtained from experimental barrier heights of corresponding *Schottky* contacts. These slope parameters and branch-point energies used as input values as well as the slope parameters, φ_v and S_{Xr}, and the branch-point energies Φ_{bpr}^p determined from the corresponding linear least-squares fits are all summarized in Table 5.1.

5.2.1 Diamond

Experimental valence-band offsets of diamond heterostructures are plotted versus the *p*-type branch-point energies of the second semiconductors in Fig. 5.10.[3] The linear least-squares fit to the data is obtained as

$$\Delta W_v - S_X(X_{\text{semi}} - X_C) = (0.90 \pm 0.27)\left[\Phi_{bp}^p(\text{semi}) - (1.81 \pm 0.81)\right][\text{eV}] \quad (5.18)$$

The resulting branch-point energy for diamond, $\Phi_{bp}^p = 1.81$ eV, although exhibiting a large margin of error, agrees with the values obtained from the experimental *Schottky* barriers of *p*-diamond analyzed in Fig. 5.1 and calculated by Mönch (1996).

---

[3] Figure 5.10. Experimental data from Cañas et al. (2021), Chen et al. (2023), Maréchal et al. (2017), Shammas et al. (2017), Shi et al. (2011a, b), Tang et al. (2011), Wang et al. (2022), Yang et al. (2020).

Fig. 5.10 Experimental valence-band offsets of diamond heterostructures minus the respective IFIGS electric-dipole contributions as a function of calculated and empirical *p*-type branch-point energies of the respective other semiconductors. The dashed line is the linear least-squares fit to the data

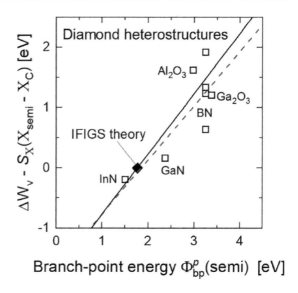

5.2.2   Silicon

Experimental barrier heights of *n*-Si *Schottky* contacts and valence-band offsets of Si heterostructures are displayed in Fig. 5.11[4] and 5.12[5], respectively. As already discussed in Sect. 3.5.3, the linear least-squares fits to the data of the Si *Schottky* contacts yield

$$\Phi_{Bn}^{(1\times1)i} = (0.803 \pm 0.005) + (0.098 \pm 0.007)(X_m - X_{Si})[eV]$$

and

$$\Phi_{Bn}^{(7\times7)i} = (0.729 \pm 0.012) + (0.092 \pm 0.008)(X_m - X_{Si})[eV]$$

---

[4] Figure 5.11: Experimental data from Blauärmel et al. (2000), Chen et al. (1993), Cohen et al. (1969), Crowell et al. (1962,1964), Jäger and Kassing (1977), Kavaliunas et al. (2021), King et al. (2015), Maeda et al. (1998), Miura et al. (1994), Morgan et al. (1996), Nur et al. (1995), Ohdomari et al. (1978), Paggel et al. (1998), Palm et al. (1993), Schmitsdorf et al. (1995), Sousa Pires et al. (1979), Turan et al. (2001), Weyers et al. (1999), Zhu et al. (2001).

[5] Figure 5.12: Experimental dada from Afanas'ev et al. (1996,2001,2002), Alay et al. (1997), Amy et al. (2004), Bersch et al. (2008), Bhat et al. (2011), Björk et al. (2010), Chaoudhary et al. (2022), Chambers et al. (2005), Chen et al. (2016), Chiam et al. (2008), Fulton et al. (2002,2006), Geogakilas et al. (1997), Gibbon et al. (2018), Grunthaner and Grunthaner [1986], Himpsel et al. (1988), Hirose et al. (2001), Jin et al. (2006), Kavaliunas et al. (2021), King et al. (2015), Keister et al. (1999), Kuo et al. (2011), Lei et al. (2012), Lucovsky et al. (2003), Margaritondo et al. (1982), Miyazaki [2001], Ohmi et al. (1988), Ohta et al. (2004), Oshima et al. (2003), Renault et al. (2004), Sayan et al. (2002), Sorifi et al. (2022), Wahab et al. (1996), Wang et al. (2004), Williams (1995), Yadav et al. (2020), Yamaoka et al. (2002), You et al. (2010), Yu et al. (2002).

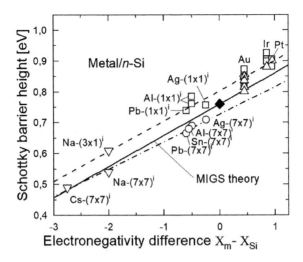

Fig. 5.11 Experimental barrier heights of laterally homogeneous *n*-type Si *Schottky* contacts versus the difference in the *Miedema* electronegativities of the metals and Si. The ○ and △, ▲, ◇, and ▽ symbols differentiate the data from I/V, BEEM, IPEYS, and PES measurements, respectively. The dashed and dash-dotted lines are the linear least-squares fits to the experimental data of diodes with $(1 \times 1)^i$ and $(7 \times 7)^I$ interface structures, respectively. The solid IFIGS line is drawn with $S_X = 0.101$ eV/Miedema-unit and $\Phi_{bp}^p = 0.36$ eV. After Mönch (2004)

Fig. 5.12 Experimental valence-band offsets of Si heterostructures minus the respective IFIGS electric-dipole contributions as a function of calculated and empirical *p*-type branch-point energies of the respective other semiconductors. The dashed line is the linear least-squares fit to the experimental data

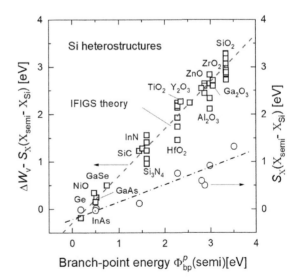

Within the margins of error, the slope parameters are identical, i.e., they do not depend on the interface structure. The difference of 73 meV between the two values of the zero-charge-transfer energies was explained by the dipole of the stacking fault of the (7x7)[i] interface structure. As already mentioned in the previous section, the p- and the n-type branch-point energies of Si, 0.38 eV and 0.80 eV, determined from the experimental data shown in Figs. 5.1 and 5.11, respectively, add up to 1.18 eV. This value is slightly larger than the bulk band gap of 1.12 eV. The linear least-squares fit to the data of Si heterostructures shown in Fig. 5.12

$$W_v - S_X(X_{semi} - X_{Si}) = (0.99 \pm 0.03)[_b p^P(semi) - (0.29 \pm 0.08)][eV] \qquad (5.19)$$

has the slope parameter $\varphi_{vbo} = 0.99 \pm 0.03$ in excellent agreement with the value demanded by Eq. (5.16). The branch-point energy $\Phi^p_{bp} = (0.29 \pm 0.03)$ eV is smaller than the result obtained from the p-Si *Schottky* contacts, but it adds up to 1.09 eV with the value of n-Si *Schottky* contacts, which corresponds to the band gap within the margins of error. These findings suggest that the branch-point energy Φ^p_{bp} of Si might be smaller than the value calculated by Tersoff (1984a) and adopted here. The respective values reported by Robertson and Falabretti (2016), Bechstedt and co-workers (Höffling et al. 2012), Hinuma et al. (2014), and Ghosh et al. (2022), which are summarized in Table 4.1, are indeed smaller than Tersoff's early result.

The circular data points in Fig. 5.12 represent the IFIGS dipole contributions $S_X(X_{semi} - X_{Si})$ to the valence-band offsets of the respective Si heterostructures. The dash-dotted line serves only as a guide for the eye. For the SiO_2-Si heterostructures, this term is estimated to be 1.32 eV, and it thus accounts for approximately a quarter of the total valence-band offset.

5.2.3   Germanium

Experimental barrier heights Φ^{hom}_{Bn} and valence-band offsets ΔW_v of Ge *Schottky* contacts and heterostructures are displayed in Figs. 5.13 and 5.14[6,7], respectively. The linear least-squares fits to the data are

---

[6] Figure 5.13: Experimental Δ, and ∇ data from *Dimulas* et al. (2006), *Nishimura* et al. (2007), and *Lieten* et al. (2008), respectively.

[7] Figure 5.14: Experimental data from Afanas'ev et al. (2011), Ban et al. (1997), Botzakaki et al. (2020), Cantoni et al. (2011), Chambers et al. (2017), Chui et al. (2005), Clavel et al. (2017), Dixit et al. (2014), Du et al. (2018), Guo et al. (2010), Hudait & Zhu (2013), Jain et al. (2014), Jeon et al. (2010), Kim et al. (2021), Kraut et al. (1980), Lim et al. (2017), Margaritondo et al. (1981), Ohta et al. (2006), Owen et al. (2013), Pa et al. (2015), Perego et al. (2006, 2007), Petti et al. (2011), Preobrajenski et al. (2000), Ruckh et al. (1994), Rumaiz et al. (2010), Seo et al. (2005), Singh et al. (2012), Spiga et al. (2005), Swaminathan et al. (2011), Wang et al. (2004, 2016), Xian et al. (2013), Xu et al. (1998), Yang et al. (1996, 2009, 2004), Yu et al. (1993), Zhang et al. (2013).

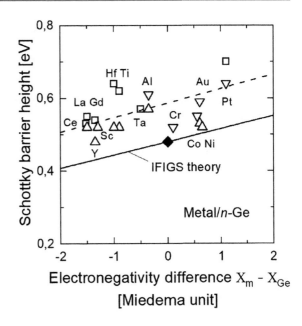

Fig. 5.13 Experimental barrier heights of laterally homogeneous *n*-type Ge *Schottky* contacts versus the difference in the *Miedema* electronegativities of the metals and Ge. The dashed line is the linear least-squares fit to the experimental data. The solid IFIGS line is drawn with $S_X = 0.04\,eV/Miedema\text{-}unit$ and $\Phi_{bp}^{p} = 0.18\,eV$

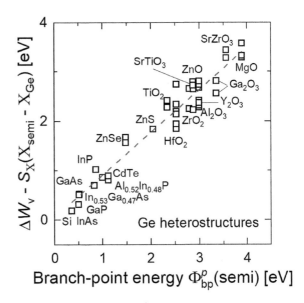

Fig. 5.14 Experimental valence-band offsets of Si heterostructures minus the respective IFIGS electric-dipole contributions as a function of calculated and empirical *p*-type branch-point energies of the respective other semiconductors. The dashed line is the linear least-squares fit] to the experimental data. From Mönch (2017)

$$\Phi_{Bn}^{hom} = (0.81 \pm 0.02) + (0.05 \pm 0.02)(X_m - X_{Ge})[eV] \qquad (5.20)$$

and

$$\Delta W_v - S_X(X_{semi} - X_{Ge}) = (0.86 \pm 0.04)[\Phi_{bp}^{p}(semi) - (0.05 \pm 0.09)][eV] \qquad (5.21)$$

respectively. Obviously, the experimental branch-point energy $\Phi_{bp}^{n} = (0.61 \pm 0.02)[\text{eV}]$ is by 0.13 eV larger than the value predicted by using Tersoff's calculated p-type branch-point energy (1984). The p-type branch-point energy $\Phi_{bp}^{p} = (0.05 \pm 0.09)[\text{eV}]$ obtained from the linear least-squares fit to the experimental valence-band offsets indeed places the IFIGS branch point of Ge close to the valence-band maximum. This finding confirms the branch-point energy $\Phi_{bp}^{p} = 0.06\text{eV}$ calculated by Brudnyi et al. (1995). Finally, the experimental Φ_{bp}^{n} and Φ_{bp}^{p} results add up to 0.66 eV which value agrees with the width of the Ge band gap at room temperature. This finding again suggests that the branch-point energy calculated by *Tersoff* is not the correct value.

5.2.4 Silicon Carbide 3C-, 4H- and 6H-SiC

Experimental barrier heights Φ_{Bn}^{hom} of n-type *Schottky* contacts and valence-band offsets ΔW_v of heterostructures of the three 3C-, 4H-, and 6H-SiC polytypes are displayed in Figs. 5.15 and 5.16[8,9], respectively. The linear least-squares fits of the 3C-, 4H-, and 6H-SiC barrier-height data result to

$$\Phi_{Bn}^{hom} = (0.92 \pm 0.10) + (0.13 \pm 0.11)(X_m - X_{SiC})[\text{eV}], \tag{5.22}$$

$$\Phi_{Bn}^{hom} = (1.63 \pm 0.03) + (0.19 \pm 0.03)(X_m - X_{SiC})[\text{eV}], \tag{5.23}$$

and

$$\Phi_{Bn}^{hom} = (1.35 \pm 0.05) + (0.17 \pm 0.04)(X_m - X_{SiC})[\text{eV}], \tag{5.24}$$

respectively. The slope parameters S_X of the three polytypes agree within the margins of error, although the value for the 3C-SiC *Schottky* contacts is slightly smaller. This observation can be easily attributed to the scattering of the barrier heights of Au and Pt contacts and to missing data for metals with smaller electronegativities. Moreover, the slope parameter $S_X = (0.23 \pm 0.05)[\text{eV/Miedemaunit}]$ determined in Fig. 5.2 for

---

[8] Figure 5.15: Experimental data from Aboelfotoh et al. (2003), Aydin et al. (2007), Constantinidis et al. (1998), Defives et al. (1999), Elsbergen (1998), Elsbergen et al. (1996), Ewing et al. (2007), Giannazzo et al. (2007), Gullu et al. (2021), Huang & Wang (2015), Im et al. (1998, 2001), Karoui et al. (2008), Khemka et al. (1998), King et al. (2015), Oder & Naredla [2022], Roccaforte et al. (2003, 2022), Sefaoglu et al. (2008), Shigiltchoff et al. (2003), Skromme et al. (2000), Suezaki et al. (2001), Toumi et al. (2021).

[9] Figure 5.16: Experimental data from Afanas'ev & Stesmans, A. (2000), Afanas'ev et al. (1996), Alivov et al. (2006), Andres et al. (2008), Chang et al. (2011), Chen et al. (2005, 2008), Choi et al. (2005), Fan et al. (2008), Gao et al. (2003), King et al. (1999), Mahapatra et al. (2008), Mattern et al. (1998), Polyakov et al. (2002), Rizzi et al. (1999), Suri et al. (2010), Tanner et al. (2007), Torvik et al. (1998), Wahab et al. (1996), Zhang et al. (2008).

Fig. 5.15 Experimental barrier heights of laterally homogeneous *n*-type SiC *Schottky* contacts versus the difference in the *Miedema* electronegativities of the metals and SiC. The dashed lines are the linear least-squares fits to the experimental data

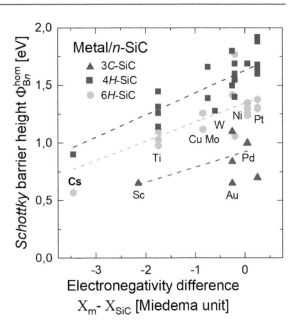

Fig. 5.16 Experimental valence-band offsets of SiC heterostructures minus the respective IFIGS electric-dipole contributions as a function of calculated and empirical, respectively, *p*-type branch-point energies of the respective other semiconductors. The dashed and dash-dotted lines are the linear least-squares fits to all the SiC heterostructure data, to the 3*C*- (▲), 4*H*- (■), and 6*H*-SiC data (●)

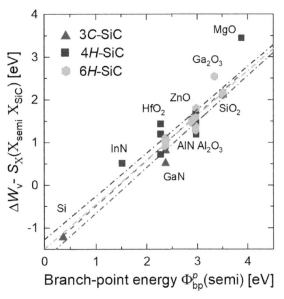

Schottky contacts on *p*-type SiC polytypes agrees with the results obtained here within the margins of error. Considering the respective band gaps of these polytypes, 3.23 and 3.00 eV, the *p*-type branch-point energies then result as 1.60 and 1.65 eV, respectively. Within the margins of experimental error these results well agree with the value of 1.58 eV obtained from the linear fit to the *p*-type data plotted in Fig. 5.2.

The valence-band offsets of the 3*C*- (▲), 4*H*- (■), and 6*H*-SiC heterostructures (●) are described by the linear least-squares fits

$$\Delta W_v - S_X(X_{semi} - X_{SiC}) = (1.07 \pm 0.07)[\Phi_{bp}^p(semi) - (1.61 \pm 0.18)][eV], \quad (5.25)$$

$$\Delta W_v - S_X(X_{semi} - X_{SiC}) = (1.01 \pm 0.18)[\Phi_{bp}^p(semi) - (1.25 \pm 0.18)][eV], \quad (5.26)$$

and

$$\Delta W_v - S_X(X_{semi} - X_{SiC}) = (1.04 \pm 0.10)[\Phi_{bp}^p(semi) - (1.43 \pm 0.20)][eV], \quad (5.27)$$

respectively. As required by Eqs. (4.15) and (4.16), the slope parameters φ_{vbo} are equal to 1 within the margins of error. Moreover, the branch-point energies Φ_{bp}^p of the three polytypes are equal within the margins of error. Therefore, all the experimental data of the three polytypes are considered together. The resulting least-squares fit to the data is found as

$$\Delta W_v - S_X(X_{semi} - X_{SiC}) = (1.07 \pm 0.06)[\Phi_{bp}^p(semi) - (1.45 \pm 0.20)][eV] \quad (5.28)$$

The common branch-point energy of the three SiC polytypes, $\Phi_{bp}^p = 1.45eV$, is slightly smaller than the value obtained in Fig. 5.2 with the *p*-SiC *Schottky* contacts, but again both values agree within the margins of error. Moreover, the value determined here excellently confirms the value calculated for 3*C*-SiC by Mönch (1996), namely $\Phi_{bp}^p = 1.44eV$.

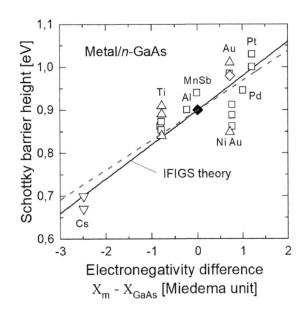

Fig. 5.17 Barrier heights of laterally homogeneous *n*-type GaAs *Schottky* contacts versus the difference in the Miedema electronegativities of the metals and GaAs. The □ and ○, ◇, and ▽ symbols differentiate the data from *I/V*, BEEM, IPEYS, and PES measurements, respectively. The dashed line is the linear least-squares fit to the experimental data. The solid IFIGS line is drawn with $S_X = 0.08$ eV/Miedema-unit and $\Phi_{bp}^{p} = 0.52$ eV. From Mönch (2017)

5.2.5   Gallium Arsenide GaAs

Figures 5.17 and 5.18[10,11] display experimental barrier heights of laterally homogeneous *n*-GaAs *Schottky* contacts and valence-band offsets of GaAs heterostructures, respectively. The dashed lines are the linear least-squares fits

$$\Phi_{Bn}^{hom} = (0.90 \pm 0.01) + (0.07 \pm 0.01)(X_m - X_{GaAs})[eV] \tag{5.29}$$

and

$$\Delta W_v - S_X(X_{semi} - X_{GaAs}) = (1.02 \pm 0.04)[\Phi_{bp}^{p}(semi) - (0.65 \pm 0.09)][eV] \tag{5.30}$$

---

[10] Figure 5.17: Experimental data from Arulkumaran et al. (1996), Bhuiyan et al. (1988), Cola et al. (1992), Dharmarasu et al. (1998), DiDio et al. (1995), Dogan et al. (2007), Grunwald [1987], Hübers & Röser (1998), Leroy et al. (2005), Manago et al. (2000), Nathan et al. (1996), Nuhoglu et al. (1998), Özdemir et al. (2006), Sehgal et al. (1998), Spicer et al. (1975), Stockmann & Kempen (1998).

[11] Figure 5.18: Experimental data from Dalapati et al. (2008}, Ding et al. (1987), Golan et al. (1991), Klein et al. (1997), Kowalczyk et al. (1982a, b), Kraut et al. (1980), Liang et al. (2005), Lu et al. (2006a), Nguyenet al. (2008), Waag et al. (1990), Wang and Stern (1985), Wang et al. (2011), Yu et al. (1990), Zhang et al. (2008).

Fig. 5.18 Experimental valence-band offsets of GaAs heterostructures minus the respective IFIGS electric-dipole contributions as a function of calculated and empirical p-type branch-point energies of the respective other semiconductors. The dashed line is the linear least-squares fit to the data. The solid IFIGS line is drawn with Φ_{bp}^{p} = 0.52 eV

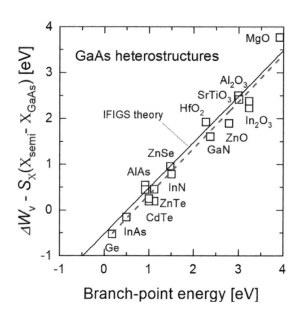

to the respective data. The slope parameter S_{Xn} = (0.07±0.01) [eV/Miedema unit] of Eq. (5.29) agrees with S_{Xp} = (0.11±0.03) [eV/Miedema unit] within the margins of error, which value is obtained from the barrier heights of p-GaAs *Schottky* contacts plotted in Fig. 5.5. As required by Eqs. (5.15) and (5.16), the slope parameter φ_{vbo} of Eq. (5.30) is equal to 1 within the margins of error. The branch-point energies Φ_{bp}^{p} = 0.65 eV and 0.64 eV, which are obtained from the experimental data of heterostructures shown in Fig. 5.18 and the p-type *Schottky* contacts displayed in Fig. 5.5, respectively, match with each other, but even within the margins of error they are larger than the value of 0.52 eV calculated by Mönch (1996). Moreover, the branch-point energy Φ_{bp}^{n} = 0.9 eV, which results from the barrier heights of the n-GaAs *Schottky* contacts plotted in Fig. 5.17, and the calculated branch-point energy Φ_{bp}^{p} = 0.52 eV add up to the width of the GaAs band gap of 1.42 eV at room temperature.

5.2.6　Gallium Nitride GaN

Experimental barrier heights of laterally homogeneous *n*-GaN *Schottky* contacts and valence-band offsets of GaN heterostructures are displayed in Figs. 5.19 and 5.20[12,13], respectively. The linear least-squares fits to the data are

$$\Phi_{Bn}^{hom} = (1.10 \pm 0.02) + (0.26 \pm 0.02)(X_m - X_{GaN})[eV] \tag{5.31}$$

and

Fig. 5.19 Barrier heights of laterally homogeneous *n*-GaN(0001) *Schottky* contacts versus the difference in the Miedema electronegativities of the metals and GaN. The *solid* IFIGS *line* is drawn with $S_X = 0.29\,eV/Miedema$-unit and $\Phi_{bp}^{p} = 2.37\,eV$. The □, ○, ▽ and △ symbols differentiate the data from *I/V*, *C/V*, IPEYS, and PES measurements. The dashed line is the linear least-squares fit $\Phi_{Bn}^{hom} = (1.10 \pm 0.02) + (0.26 \pm 0.02)(X_m - X_{GaN})[eV]$ to the data

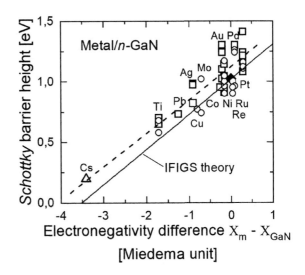

---

[12] Figure 5.19: Experimental data from Allen et al. (2020), Bell et al. (1998), Binari et al. (1994), Dogan and Elagoz (2014), Ejderha et al. (2014), Eyckeler et al. (1998), Hacke et al. (1993), Helal et al. (2020), Iucolano et al. (2007), Jyothi et al. (2010), Kalinina et al. (1997), Kampen and Mönch (1997), Karrer et al. (1999, 2000), Khan et al. (1995), Kim (2010, 2024), Kribes et al. (1997), Lin (2009), Liu et al. (1998), Makimoto et al. (2003), Mamor (2009), Mohammad et al. (1996), Molina and Mohney (2022), Oswald et al. (2005), Ozbek and Baliga (2011), Ping et al. (1996), Qiao et el. (2000), Reddy et al. (2006, 2008, 2011,2012), Roccaforte et al. (2019), Roul et al. (2012), Sarpatwari et al. (2011), Sawada et al. (2000), Schmitz et al. (1998), Wang and He (1998), Wang et al. (1996), Yu et al. (1998), Zhou et al. (2007).

[13] Figure 5.20: Experimental data from Alivov et al. (2008), Chen et al. (2005,2006), Cook et al. (2003a,b,c), Craft et al. (2007), Das et al. (2020), Ding et al. (1997), Duan et al. (2020), Grodzicki et al. (2014), Hansen et al. (2005), Hong et al. (2001), Jia et al. (2017), Kim (2010), King et al. (1998, 1999,2007,2008), Lee et al. (2008), Lewandków et al. (2021), Lin (2009), Liu et al. (2008,2011), Maffeis et al. (2000), Mahmood et al. (2007), Makimoto et al. (2002), Martin et al. (1994,1996), Miyake et al. (2009), Ohashi et al. (2006), Polyakov et al. (2002), Rizzi et al. (1999), Shi et al. (2011), Shih et al. (2005), Supardan et al. (2020), Torvik et al. (1998), Wan et al. (2013), Wei et al. (2012), Wu et al. (2007, 2008).

$$\Delta W_v - S_X(X_{semi} - X_{GaN}) = (0.91 \pm 0.03)[\Phi_{bp}^{p}(semi) - (2.32 \pm 0.08)][eV] \quad (5.32)$$

respectively. The difference between the slope parameters $S_{Xn} = (0.26 \pm 0.02)$ [eV/Miedema unit] and $S_{Xp} = (0.10 \pm 0.11)$ [eV/Miedema unit] of the n- and p-GaN Schottky contacts displayed in Figs. 5.19 and 5.3, respectively, is most probably due to the small number of data for the p-type contacts and their large scatter. The excellent agreement of the branch-point energies $\Phi_{bp}^{p} = 2.32$ eV and 2.30 eV, which are obtained from the experimental data of heterostructures shown in Fig. 5.19 and the p-type Schottky contacts displayed in Fig. 5.3, respectively, support the previous conjecture. The slope parameter $\varphi_{vbo} = (0.91 \pm 0.03)$ of the fit (5.32) is slightly smaller than the value 1 required by Eqs. (5.15) and (5.16). Finally, the branch-point energies $\Phi_{bp}^{n} = 1.1$ eV and $\Phi_{bp}^{p} = 2.32$ eV obtained from fits (5.31) and (5.32), respectively, add up to 3.42 eV, which is the width of the GaN band gap at room temperature.

5.2.7 Aluminum Nitride AlN

The few barrier heights of n-AlN Schottky contacts plotted versus the electronegativity difference $X_m - X_{AlN}$ in Fig. 5.21[14] scatter very widely. Reddy et al. (2014) suggested that there exists a distinct offset of 0.6 eV between barrier heights of contacts on polar and non-polar AlN substrates. Because of the large scatter of the data this conclusion appears not convincing. A tentative linear least-squares fit to all the data results in

$$\Phi_{Bn}^{hom} = (2.18 \pm 0.11) + (0.31 \pm 0.10)(X_m - X_{AlN})[eV] \quad (5.33)$$

Fig. 5.20 Experimental valence-band offsets of GaN heterostructures minus the respective IFIGS electric-dipole contributions as a function of the calculated and empirical p-type branch-point energies of the respective other semiconductors. The dashed line is the linear least-squares fit to the experimental data

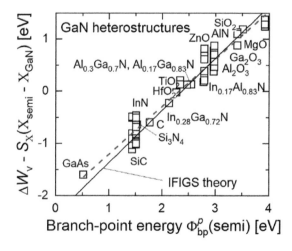

[14] Figure 5.21: Experimental data from Fu et al. (2017), Ran et al. (2020), Reddy et al. (2014), Zhou et al. (2019).

Fig. 5.21 Barrier heights of laterally homogeneous *n*-AlN *Schottky* contacts versus the difference in the Miedema electronegativities of the metals and AlN. The dashed line is the linear least-squares fit to the experimental data

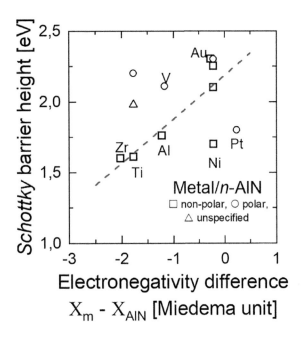

The valence-band offsets of AlN heterostructures displayed in Fig. 5.22[15] indicate that, with the exception of Ga_2O_3- and MgO/AlN heterostructures, the valence-band offsets of all other AlN heterostructures have negative signs. Therefore, the relation (5.16) can be used to determine the branch-point energy Φ_{bp}^p and the slope parameter S_X of AlN. In Fig. 5.23[13] the negative valence-band offsets of AlN heterostructures minus the branch-point-energies of the respective other semiconductors are plotted versus the difference $X_{semi} - X_{AlN}$ of the electronegativities of the second semiconductor and AlN. The linear least-squares fit

$$\Delta W_v - \Phi_{bp}^p(semi) = -(3.04 \pm 0.05) + (0.36 \pm 0.47)(X_{semi} - X_{AlN})[eV] \qquad (5.34)$$

gives the branch-point energy $\Phi_{bp}^p = 3.04 eV$ and the slope parameter $S_X = 0.36$ eV/ Miedema unit. Within the large margins of error, the two values of the slope parameter obtained from the fits (5.33) and (5.34) are compatible. The latter S_X value is now used in Fig 5.22 where the differences $\Delta W_v - S_X(X_{semi} - X_{AlN})$ are plotted against the branch-point energy $\Phi_{bp}^p(semi)$ of the second semiconductor. The linear least-squares fit to the data is

---

[15] Figures 5.22 and 5.23: Experimental data from Chen et al. (2017), Choi et al. (2005), Du et al. (2018), Fujii et al. (2009), King et al. (1998, 2015, 2007), Kuo et al. (2011), Li et al. (2014), Liu et al. (2010), Martin et al. (1994, 1996), Mietze et al. (2011), Rizzi et al. (1999), Sun et al. (2017a, b), Veal et al. (2008), Yang et al. (2009), Zhao et al. (2018).

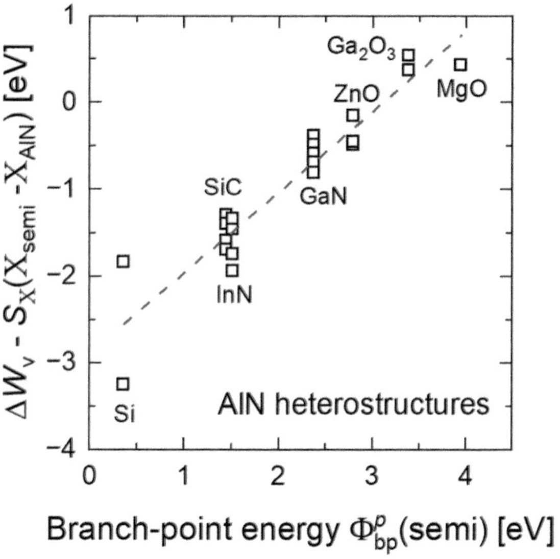

Fig. 5.22 Experimental valence-band offsets of AlN heterostructures minus the respective IFIGS electric-dipole contributions as a function of the calculated and empirical p-type branch-point energies of the respective other semiconductors and insulators. The dashed line is the linear least-squares fit to the experimental data

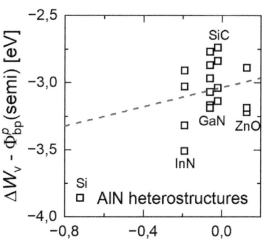

Fig. 5.23 Experimental valence-band offsets of AlN heterostructures minus the p-type branch-point energies of the other semiconductor as a function of the electronegativity difference $X_{semi} - X_{AlN}$. The dashed line is the linear least-squares fit to the experimental data

Fig. 5.24 Experimental valence-band offsets of InN heterostructures minus the respective IFIGS electric-dipole contributions as a function of the calculated and empirical *p*-type branch-point energies of the respective other semiconductors. The dashed line is the linear least-squares fit to the experimental data

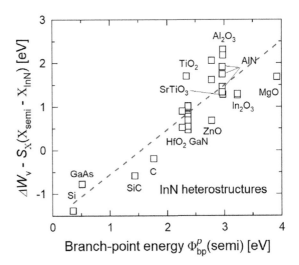

$$\Delta W_v - S_X(X_{semi} - X_{AlN}) = (0.93 \pm 0.07)\left[\Phi_{bp}^P(semi) - (3.10 \pm 0.15)\right][eV] \quad (5.35)$$

Within the margins of error, the slope parameter $\varphi_{vbo} = (0.93 \pm 0.07)$ agrees with the value 1 predicted in relation (5.16).

The two branch-point energies, 3.04 eV and 3.10 eV, resulting from the fits (5.25) and (5.26), respectively, are the same within the margins of error. Mönch (1996) and later Schleife et al. (2009) and Belabbes et al. (2011) calculated the *p*-type branch-point energy of AlN to be 2.97 eV and 3.35 eV, respectively. These theoretical values compare well with the experimental results obtained from the fits (5.34) and (5.35) to the measured valence-band offsets.

5.2.8 Indium Nitride InN

Experimental valence-band offsets of InN heterostructures minus the respective IFIGS electric-dipole contributions are displayed in Fig. 5.24.[16] The linear least-squares fit to the data is

$$\Delta W_v - S_X(X_{semi} - X_{InN}) = (1.02 \pm 0.11)[\Phi_{bp}^P(semi) - (1.54 \pm 0.28)][eV] \quad (5.36)$$

---

[16] Figure 5.24: Experimental data from Eisenhardt et al. (2010, 2012), Fujii et al. (2009), Jia et al. (2015), King et al. (2007, 2008), Krishna & Gupta (2014), Kuo et al. (2011), Li et al. (2011), Mahmood et al. (2007), Martin et al. (1996), Ohashi et al. (2006), Shi et al. (2010), Shih et al. (2005, 2008), Song et al. (2009), Wang et al. (2007), Wu et al. (2007, 2008), Yang et al. (2009), Zhang et al. (2007, 2008).

The slope parameter $\varphi_{vbo} = (1.02 \pm 0.11)$ of the fit (5.36) is very close to the value 1 required by Eqs. (5.15) and (5.16). The branch-point energy $\Phi_{bp}^{p} = 1.54$eV excellently confirms the theoretical values of 1.51 and 1.58 eV calculated by Mönch (1996) and Schleife et al. (2009), respectively.

Unfortunately, there are still no detailed studies on InN *Schottky* contacts. The probably essential reason for this is that the investigated metal-InN contacts at all show ohmic behavior. Rickert et al. (2003) did use XPS to determine the distance of the valence band top to the *Fermi* level at Ti- and Au/InN interfaces. However, the InN surfaces were differently prepared before the metal depositions. Nevertheless, Rickert et al. concluded that *"if the band gap value of 0.7 eV is employed, then E_F would be at or in the conduction band. This Fermi level position would be consistent with the observed ease of Ohmic contact formation."* At that time, the width of the InN band gap was not precisely known. Furthermore, Rickert et al. did not realize that calculations had placed the InN branch point in the conduction band (Mönch 1996).

5.2.9 Gallium Oxide Ga₂O₃

The experimental barrier heights of n-Ga$_2$O$_3$ *Schottky* contacts, plotted in Fig. 5.25[17], exhibit a very large scatter. For example, the values reported for Ir and Au contacts vary between 1.18 and 2.25 eV and 1.07 and 1.98 eV, respectively, i.e., by 1 eV. Using density-functional theory (DFT), Therrrien et al. (2021) predicted barrier heights of about 2 eV for clean Pt/n-β-Ga$_2$O$_3$(201) contacts. They attributed the much lower barrier heights reported by other authors to *"water products"* at the interface, i.e., residues of chemical treatments prior to the metal deposition. Looking first at the experimental data of the Ga$_2$O$_3$ heterostructures will reveal a simple way out of this dilemma.

---

[17] Figure 5.25: Experimental data from Ahn et al. (2017), Altuntas et al. (2014), Buzio et al. (2020), Fang et al. (2022), Fares et al. (2019), Farzana et al. (2017), Feng et al. (2018), Fu et al. (2018), Harada et al. (2019), He et al. (2017), Higashiwaki et al. (2015), Hou et al. (2019a, b, c), Hu et al. (2019), Irmscher et al. (2011), Jian et al. (2018, 2020), Joishi et al. (2018), Lingaparthi et al. (2020), Lyle et al. (2021), Mohamed et al. (2012), Oishi et al. (2015), Sheoran et al. (2020), Splith et al. (2014), Suzuki et al. (2009), Xu et al. (2019), Yang et al. (2017, 2018a, b, 2019), Yao et al. (2017), Zhang et al. (2016), Zhang et al. (2019).

Fig. 5.25 Barrier heights of laterally homogeneous *n*-Ga$_2$O$_3$ *Schottky* contacts versus the difference in the Miedema electronegativities of the metals and Ga$_2$O$_3$. The solid IFIGS "concept" line is drawn with $S_X = 0.38$ eV/Miedema-unit and $\Phi_{bp}^n = (4.85\text{-}3.35)$ eV $= 1.510$ eV. The dashed line is the linear least-squares fit to the data. The dash-dotted line is only meant to guide the eye

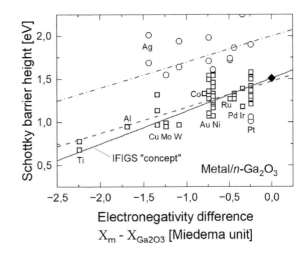

The data presented in Fig. 5.26[18] show that the valence-band offsets of Ga$_2$O$_3$ heterostructures have negative sign, with SiO$_2$- and MgO-Ga$_2$O$_3$ being exceptions. This means that Eq. (5.5) can be applied to the ΔW_v data of Ga$_2$O$_3$. Consequently, the valence-band offsets of the Ga$_2$O$_3$ heterostructures minus the branch-point energies Φ_{bpl}^p of the other semiconductors are plotted in Fig. 5.27[18] versus the difference $X_{sl} - X_{sr}$ of the electronegativities of the second semiconductors and of Ga$_2$O$_3$. Naturally, the data of the SiO$_2$- and MgO-Ga$_2$O$_3$ heterostructures are not considered here. The linear least-squares fit

$$\Delta W_v - \Phi_{bp}^p(\text{semi}) = -(3.35 \pm 0.09) + (0.34 \pm 0.15)(X_{\text{semi}} - X_{\text{Ga2O3}})[\text{eV}] \quad (5.37)$$

gives the slope parameter $S_X = (0.34 \pm 0.15)$ [eV/Miedema unit] and the branch-point energy $\Phi_{bp}^p = (3.35 \pm 0.09)$ [eV] of Ga$_2$O$_3$. Robertson and Falabretti (2006) calculated the branch-point energy of Ga$_2$O$_3$ as 2.8 eV which is slightly smaller than the value resulting from the experimental data.

The solid line in Fig. 5.25 is drawn with $S_X = 0.34$ eV/Miedema-unit and $\Phi_{bp}^n = (4.85\text{-}3.35)$ eV $= 1.50$ eV, the data obtained from the linear least-squares fit (5.37). Obviously,

---

[18] Figures 5.26 and 5.27: Experimental data from Carey et al. (2017)), Chang et al. (2011), Chen et al. (2016), Chen et al. (2018), Chen et al. (2019), Chen et al. (2022), Deng et al. (2023), Fares et al. (2019), Ghosh et al. (2019), Gibbon et al. (2018), Grodzicki et al. (2014), Hattori et al. (2016), Hu et al. (2020), Huan et al. (2018), Ji et al. (2021), Jia et al. (2015), Kamimura et al. (2014), Kuang et al. (2021), Konishi et al. (2016), Li et al. (2019), Li et al. (2020a, b), Liu et al. (2019), Liu et al. (2020, 2021), Lu et al. (2020), Lu et al. (2023), Matsuo et al. (2018), Oshima et al. (2018), Petkov et al. (2023), Schultz et al. (2020), Sun et al. (2017), Sun et al. (2018a, b), Wei et al. (2012), Wheeler et al. (2017), Yadav et al. (2020), Yamamoto et al. (2018), Yang et al. (2019, 2024), Yuan et al. (2018), Zhang et al. (2018), Zhang et al. (2020a, b), Zhi et al. (2020).

Fig. 5.26 Experimental valence-band offsets ΔW_v of Ga_2O_3 heterostructures as a function of the calculated and empirical p-type branch-point energies Φ_{bp}^p of the respective other semiconductors. The dashed line is only meant to guide the eye

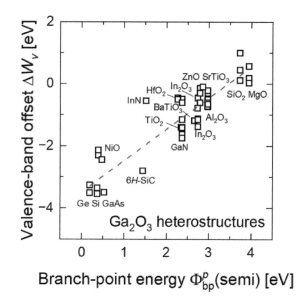

Fig. 5.27 Experimental valence-band offsets of Ga_2O_3 heterostructures minus the p-type branch-point energies of the other semiconductors as a function of the electronegativity difference $X_{semi} - X_{Ga2O3}$

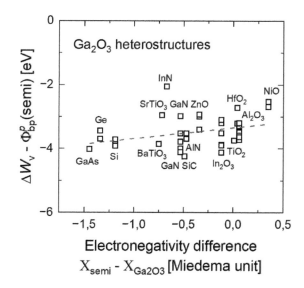

it excellently explains the square data points which are characterized by the linear least-squares fit

$$\Phi_{Bn}^{hom} = (1.47 \pm 0.03) + (0.30 \pm 0.04)(X_m - X_{Ga2O3}) \tag{5.38}$$

The experimental barrier heights represented by circular data points were not considered. The deviations from the general trend are presumably caused by extrinsic interface

Fig. 5.28 Barrier heights of laterally homogeneous Al₂O₃ *Schottky* contacts versus the difference in the *Miedema* electronegativities of the metals and Al₂O₃. The dashed line is the linear least-squares fit to the experimental data

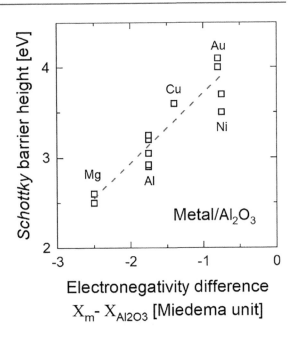

dipoles of unknown origin. Here, the residual "water products" proposed and discussed by Therrrien et al. (2021), or even thicker interfacial layers, might come into play.

5.2.10 Aluminum Oxide Al₂O₃

Figures 5.28[19] and 5.29[20] display experimental barrier heights of laterally homogeneous Al₂O₃ *Schottky* contacts and valence-band offsets of Al₂O₃ heterostructures, respectively. The dashed lines are the linear least-squares fits

$$\Phi_{Bn}^{hom} = (4.50 \pm 0.15) + (0.78 \pm 0.10)(X_m - X_{Al2O3})[eV] \tag{5.39}$$

and

$$\Delta W_v - S_X(X_{semi} - X_{Al2O3}) = (0.97 \pm 0.04)\left[\Phi_{bp}^p(semi) - (2.85 \pm 0.07)\right][eV] \tag{5.40}$$

---

[19] Figure 5.28: Experimental data from Afana'sev et al. (2002b), DiMaria (1974), Szydlo & Poirer (1971), Xu et al. (2013).

[20] Figure 5.29: Experimental data from Afanas'ev et al. (2001), Bhuiyan et al. (2021), Gao et al. (2003), Gong et al. (2011), Hashizume et al. (2003), Hattori et al. (2016), Huang et al. (2006), Kamimura et al. (2014), Lin et al. (2011), Liu et al. (2019), Ma et al. (2019), Nguyen et al. (2008), Oshima et al. (2018), Perego et al. (2008), Suri et al. (2010), Tanner et al. (2007), Xiang et al. (2013), Yang et al. (2012), Yang et al. (2019), Yu et al. (2002), Yuan et al. (2018).

Fig. 5.29 Experimental valence-band offsets of Al_2O_3 heterostructures minus the respective IFIGS electric-dipole contributions as a function of the calculated and empirical *p*-type branch-point energies of the respective other semiconductors. The dashed line is the linear least-squares fit to the experimental data

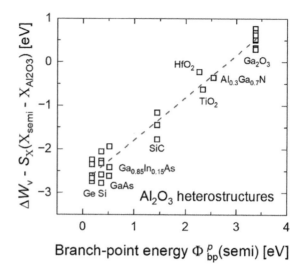

to the data. The slope parameter $\varphi_{vbo} = (0.97 \pm 0.04)$ of the fit (5.40) agrees with the value 1 required by Eqs. (5.15) and (5.16). Robertson and co-workers (2002, 2006) calculated the *p*-type branch-point energy of Al_2O_3 to be 5.5.and 6 eV, respectively. Both values are about a factor of 2 larger than the result obtained from the experimental valence-band offsets plotted in Fig. 5.29.

5.2.11 Silicon Dioxide SiO$_2$

Experimental barrier heights of SiO_2 *Schottky* contacts and valence-band offsets of SiO_2 heterostructures are shown in Figs. 5.30[21] and 5.31[22], respectively. The corresponding linear least-squares fits are

$$\Phi_{Bn}^{hom} = (4.90 \pm 0.09) + (0.75 \pm 0.09)(X_m - X_{SiO2})[eV] \qquad (5.40)$$

and

---

[21] Figure 5.30: Experimental data from Afanas'ev et al. (2002), Buchanan et al. (1998), Deal et al. (1966), Goodman and O'Neill (1966), Misra et al. (2001), Powell & Beairsto (1973).

[22] Figures 5.31 and 5.32: Experimental data from Afanas'ev & Stesmans (2000), Afanas'ev et al. (1996, 2001), Alay & Hirose (1997), Ban et al. (1998), Cook jr. et al. (2003), Fulton et al. (2006), Grunthaner & Grunthaner (1986), Himpsel et al. (1988), Hirose et al. (2001), Keiste et al. (1999), Liu et al. (2013), Lucovsky et al. (2003), Mattern et al. (1998), Miyazaki (2001), Ohta et al. (2004), Petkov et al. (2023), Renault et al. (2004), Sayan et al. (2002), Wang et al. (2013), Watanabe et al. (2011), You et al. (2009).

Fig. 5.30 Barrier heights of laterally homogeneous SiO$_2$ *Schottky* contacts versus the difference in the Miedema electronegativities of the metals and SiO$_2$. The dashed line is the linear least-squares fit to the experimental data

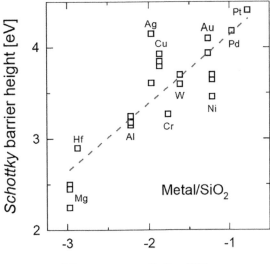

Fig. 5.31 Experimental valence-band offsets of SiO$_2$ heterostructures minus the respective IFIGS electric-dipole contributions as a function of calculated and empirical *p*-type branch-point energies of the respective other semiconductors. The dashed line is the linear least-squares fit Δ to the experimental data

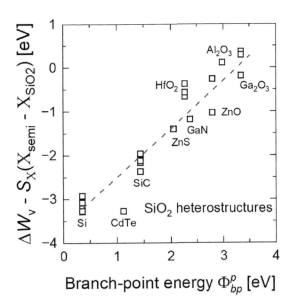

Fig. 5.32 Experimental valence-band offsets of SiO_2 heterostructures minus the p-type branch-point energies of the other semiconductor as a function of the electronegativity difference $X_{semi} - X_{SiO2}$. The dashed line is the linear least-squares fit to the data points

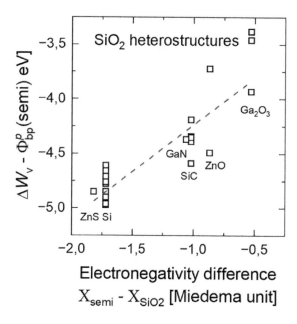

$$\Delta W_v - S_X(X_{semi} - X_{SiO2}) = (1.08 \pm 0.05)\left[\Phi_{bp}^p(semi) - (3.33 + 0.08)\right][eV], \quad (5.41)$$

respectively. The slope parameter $\varphi_{vbo} = (1.08 \pm 0.05)$ of the fit (5.41) is close to the value 1 required by Eqs. (5.15) and (5.16).

Since the p-type branch-point energies of the second semiconductors considered are all smaller than the value of SiO_2, relation (5.16) can be applied for a cross-check. Therefore, in Fig. 5.32[20], the valence-band offsets used in Fig. 5.31 minus the p-type branch-point energies of the corresponding other semiconductors are plotted as a function of the electronegativity difference $X_{semi} - X_{SiO2}$. The linear least-squares fit

$$\Delta W_v - \Phi_{bp}^p(semi) = -(3.38 \pm 0.11) + (0.86 \pm 0.08)(X_{semi} - X_{SiO2})[eV] \quad (5.42)$$

to the data points gives the slope parameter $S_X = (0.86 \pm 0.08)$ [eV/Miedema unit] and the branch-point energy $\Phi_{bp}^p = (3.38 \pm 0.11)$ [eV] of SiO_2. Within the margins of error, both values agree with the results obtained from the fits in Figs. 5.30 and 5.31.

Robertson and Falabretti (2006) and Höffling et al. (2012) calculated the p-type branch-point energy of SiO_2 as 4.5 and 4.52 eV, respectively. These values are larger than the result obtained here from the evaluation of the experimental valence-band offsets. The branch-point energies $\Phi_{bn}^p = 4.9eV$ and $\Phi_{bp}^p = 3.38eV$ add up to 8.28 eV. This value is smaller than the experimental value of the band gap of SiO_2 films that is reported to vary between 8.6 and 9.1 eV.

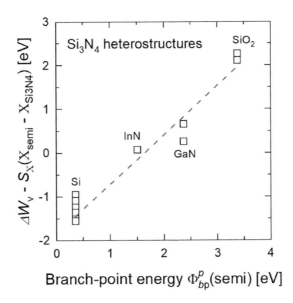

Fig. 5.33 Experimental valence-band offsets of Si_3N_4 heterostructures minus the respective IFIGS electric-dipole contributions as a function of calculated and empirical p-type branch-point energies of the respective other semiconductors. The dashed line is the linear least-squares fit to the experimental data

5.2.12 Silicon Nitride Si_3N_4

Experimental valence-band offsets of Si_3N_4 heterostructures are plotted versus the p-type branch-point energies of the other semiconductors in Fig. 5.33.[23] The linear least-squares fit to the data gives

$$\Delta W_v - S_X(X_{semi} - X_{Si3N4}) = (1.13 \pm 0.10)\left[\Phi_{bp}^p(semi) - (1.64 + 0.18)\right][eV] \quad (5.43)$$

The slope parameter $\varphi_{vbo} = (1.18 \pm 0.10)$ of this fit is slightly larger than the value 1 required by Eqs. (5.15) and (5.16). For Si_3N_4, Robertson (2013) calculated the branch point by 3 eV above the valence-band maximum. This value is much larger than the result $\Phi_{bp}^p = 1.64eV$ obtained from the fit to the experimental data in Fig. 5.33.

---

[23] Figure 5.33: Experimental data from Cook Jr. et al. (2003), DiMaria & Arnett (1975), Goodman [1968], Gritsenko et al. (2003), Higuchi et al. (2007), Keister et al. (1999), Kumar et al. (2012), Miyazaki et al. (2003).

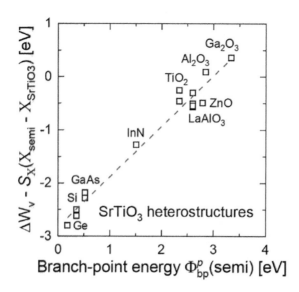

Fig. 5.34 Experimental valence-band offsets of SrTiO₃ heterostructures minus the respective IFIGS electric-dipole contributions as a function of the calculated and empirical p-type branch-point energies of the respective other semiconductors and insulators. The dashed line is the linear least-squares fit to the data points

5.2.13 Strontium Titanate SrTiO₃

Experimental valence-band offsets of $SrTiO_3$ heterostructures are plotted as a function of the p-type branch-point energies of the other semiconductors in Fig. 3.34.[24] The linear least-squares fit to the data is

$$\Delta W_v - S_X(X_{semi} - X_{SrTiO3}) = (0.95 \pm 0.04)\left[\Phi_{bp}^p(semi) - (2.82 \pm 0.10)\right]. \text{ eV} \quad (5.44)$$

The slope parameter $\varphi_{vbo} = (0.95 \pm 0.04)$ of this fit is within the margins of error equal to the value 1 required by Eqs. (5.15) and (5.16). The branch-point energy of 2.82 eV is slightly larger than the value of 2.3 eV calculated by Robertson and Falabretti (2006) (Table 5.2).

---

[24] Figure 5.34: Experimental data from Amy et al. (2004), Chambers et al. (2001, 2004, 2009, 2017b), Jia et al. (2009), Li et al. (2011), Liang et al. (2005), Lu et al. (2023), Schütz et al. (2015), Tuan et al. (2003), Yang et al. (2013).

Table 5.2 Band gap W_g in eV, optical dielectric constant ε_∞, experimental slope parameters S_X in eV/Miedema unit and branch-point energies Φ_{bp}^n and Φ_{bp}^p in eV

	W_g	ε_∞	*Schottky* contacts		Heterostructures		Reference
			S_X	Φ_{bp}^n	φ_{vbo}	Φ_{bp}^p	
GaSb	0.73	14.44	0.02	0.55	–	–	Mönch unpublished
MgO	7.8	3.00	–	–	0.99	3.92	Mönch (2011)
MgO			0.59[a]	4.24[b]	–	–	Lu et al. (2006b)
ZnO	3.37	4.03	–	–	0.98	2.79	Mönch (2011)
In$_2$O$_3$	3.71	4.5	–	–	0.83	3.23	Mönch (2011)
Gd$_2$O$_3$		3.24	–	–	0.95	2.85	Mönch (2016)
TiO$_2$	3.5	7.8	0.12	1.13	0.95	2.34	Mönch (2010)
ZrO$_2$	5.6	4.8	0.47	3.64	0.84	2.87	Mönch (2007)
HfO$_2$	5.4	4	0.66	3.96	0.91	2.31	Mönch (2007)
NiO	3.7		–	–	0.84	0.41	Mönch unpublished
SrTiO$_3$	3.25	5.56	–	–	1.0	2.86	Mönch (2018)
LaAlO$_3$	6.4	4	–	–	1.13	2.59	Mönch (2018)
InGaZnO$_4$	3.2	3.6	–	–	0.92	2.37	Mönch (2018)
LiF	–	2	1.59				Pong & Paudyal (1981)
BaF2	–	2.1	1.53				Pong & Paudyal (1981)
Xe	–	2	1.53				Jacob et al. (1987)

[a] S_X^p

[b] Φ_{bp}^p

Irradiation- or Native-Defect-Induced Gap States

6

The concept of interface-induced gap states is independently supported from an entirely different field, namely native defects in the bulk of semiconductors which are generated by irradiation with energetic particles. Lark-Horovitz and coworkers were the first to find that both electron and hole traps associated with native defects are generated by such bombardments in *n*- and *p*-type Ge (Lark-Horovitz et al. 1948) and Si (Johnson & Lark-Horovitz 1948). These traps change the resistivity of the semiconductors. With increasing irradiation time, the *Fermi* levels shift to the same *saturation* position $\Phi_{lim}^{p} = W_F^{lim} - W_v$ irrespective of whether the semiconductors are initially doped *p*- or *n*-type. Experimentally observed values are summarized in Table 6.1. Walukiewicz (1987) found *"a correlation between the Fermi level pinning deduced from Schottky barrier heights and from electrical properties of irradiated III-V semiconductors"*. His conclusion that this *"correlation indicates that similar defects are responsible for the Fermi level stabilization in both cases"* is by no means stringent. But at that time Spicer's (1979) unified defect model still dominated the respective discussions. However, no experimental evidence has been found that native defects determine the barrier heights of *Schottky* contacts (Vrijmoeth et al. 1990, Hong et al. 1992, Howes et al. 1995).

Brudnyi et al. (1995) have already mentioned that the saturation positions of the *Fermi* level due to the native defects in heavily irradiated semiconductors are connected to the branch points of the complex band structure. This statement is obvious because the gap states induced by the native defects generated by the energetic particles also originate from the complex band structure just like the interface-induced gap states at *Schottky*

© The Author(s), under exclusive license to Springer Nature Switzerland AG 2024
W. Mönch, *Electronic Structure of Semiconductor Interfaces*, Synthesis Lectures on Engineering, Science, and Technology, https://doi.org/10.1007/978-3-031-59064-1_6

Table 6.1 Experimental *Fermi*-level stabilization energies of semiconductors after prolonged irradiation with high-energy particles

	Φ_{lim}^{p} [eV]	Reference
Si	0.39	Konozenkao et al. (1969)
Si	0.48	Lugakov et al. (1982)
Ge	0.22	Bohlke et al. (1966)
Ge	0.07	Gerasimov (1978)
Ge	0.13	Brudnyi & Grinyaev (1998)
AlSb	0.50	Brudnyi & Grinyaev (1998)
GaN	2.54	Polyakov et al. (2007)
GaN	2.69	Li et al. (2005)
GaN	2.49	Brudnyi et al. (2012)
GaP	0.90	Brudnyi et al. (1985)
GaP	1.13	Novikov (1994)
GaAs	0.58	Wohlleben & Beck (1966)
GaAs	0.70	Dyment et al. (1973)
GaAs	0.60	Brudnyi et al. (1982)
GaSb	0.16	Cleland & Crawford (1955)
GaSb	0.035	Brudnyi & Grinyaev (1998)
GaSb	0.02	Boiko et al. (2015)
InN	1.6	Li et al. (2005)
InP	0.95	Donnelly & Hurwitz (1977)
InP	1.00	Brudnyi et al. (2005)
InAs	0.52	Brudnyi et al. (2003)
InSb	0.06	Mashovets & Khansevarov (1966)
InSb	0.18	Foyt et al. (1970)
InSb	0.07	Myhra (1978)
InSb	0	Brudnyi & Grinyaev (1998)
CdO	2.55	King et al. (2009)

contacts and heterostructures. Therefore, the experimentally observed *Fermi*-level saturation energies[1] Φ_{lim}^{p} are plotted versus the calculated and the experimental branch-point energies in Figs. 6.1 and 6.2, respectively. The dashed lines are the linear least-squares fits

---

[1] Figures 6.1 and 6.2: Experimental data from Bohlke et al. (1966), Boiko et al. (2015), Brudnyi & Grinyaev (1998), Brudnyi et al. (1982, 2003, 2005, 2012), Cleland & Crawford (1955), Donnelly & Hurwitz (1977), Dyment et al. (1973), Foyt et al. (1970), Gerasimov (1978), King et al. (2009), Konozenkao et al. (1969), Li et al. (2005), Lugakov et al. (1982), Mashovets & Khansevarov (1966), Myhra (1978), Novikov (1994), Polyakov et al. (2007), Wohlleben & Beck (1966).

Fig. 6.1 Experimental *Fermi*-level stabilization energies versus *p*-type branch-point energies calculated by Tersoff (1984) for Si and Ge, by Mönch (1996) for the III-V compounds and by Schleife et al. (2009) for CdO

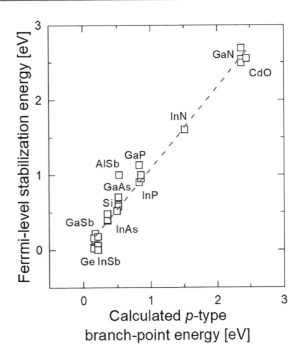

$$\Phi_{lim}^{p} = (1.08 \pm 0.02)\,\Phi_{bp}^{p}(\text{calc.})[\text{eV}] \qquad (6.1)$$

and

$$\Phi_{lim}^{p} = (1.10 \pm 0.03)\,\Phi_{bp}^{p}(\text{exp.})[\text{eV}] \qquad (6.2)$$

respectively. Within the error margins of both experimental and calculated data, the *Fermi*-level saturation energies agree with the branch-point energies.

Deep Level Transient Spectroscopy (DLTS), developed by Lang [1974], has become the standard method for determining the energy positions of traps in the band gap of irradiated semiconductors. Corresponding data of experimental investigations with SiC, GaAs, and Ga_2O_3 are shown in Figs. 6.3, 6.4 and 6.5. Experimentally observed levels of electron and hole traps induced by irradiation with energetic particles in the three most important polytypes 4*H*-, 6*H*-and 3*C*-SiC are shown in Fig. 6.3.[2] The hole and the electron traps are distinguished by red and blue dash-dotted lines, respectively. There

---

[2] Figure 6.3. Experimental data of irradiated 4*H*-, 6*H*-, and 3*C*-SiC polytypes from Alfieri & Kimoto (2009, 2011), Alfieri et al. (2016), Dalibor et al. (1997), Danno & Kimoto (2006, 2007a, b), Davydov et al. (2001), Gong et al. (1999), Hemmingson et al. (1997), Izzo et al. (2009), Kato et al. (2001), Kawahara et al. (2009), Lebedev et al. (2000a, b, c), Nagesh et al. (1990), Paradzahet al. (2015), Sasaki et al. (2011), Storasta et al. (2004), Weidner et al. (2001), Zekentes et al. (1995), Zhang et al. (2003), Zhou et al. (1987).

Fig. 6.2 Experimental
Fermi-level stabilization
energies versus experimental
p-type branch-point energies
from Table 5.1, last column

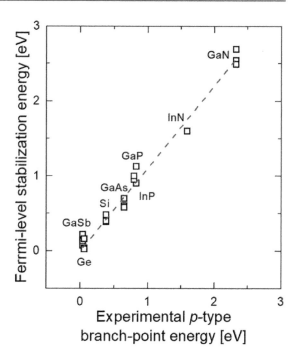

are many experimental data for the 4*H* and 6*H* polytypes. However, Alfieri and Kimoto (2009) already noted that for 3*C*-SiC very few reports on deep levels can be found in the literature (Alfieri & Kimoto 2009, Kato et al. 2001, Nagesh et al. 1990, Zekentes et al. 1995, Zhou et al. 1987) and even less on irradiation-induced traps (Nagesh et al. 1990, Alfieri & Kimoto 2009). As to be expected, the irradiation-induced electron and hole traps are located above and below the branch-point $W_{bp} = W_v + 1.44\text{eV}$, respectively. This latter value was calculated by Mönch [1996]. Furthermore, the analysis of the valence-band offsets in SiC heterostructures shown in Fig. 5.16 revealed that the branch points of the three polytypes 4*H*-, 6*H*-, and 3*C*-SiC have this same position in the gap. The data shown in Fig. 6.3 are consistent with this finding.

The few available values for irradiation-induced traps in GaAs are shown in Fig. 6.4. The hole and the electron traps are again shown by red and blue dash-dotted lines. These irradiation-induced electron and hole traps are located above and below the branch-point $W_{bp} = W_v + 0.52$ eV calculated by Mönch (1996). Within the margins of experimental error, this also applies to the electron trap E5.

Finally, Figure 6.5 shows the energy levels of hole and electron traps induced by high-energy particles in Ga_2O_3 as compiled by Kim et al. [2019]. The branch-point level $W_{bp} = W_v + (3.35 \pm 0.09)\text{eV}$ was obtained from an analysis of the valence-band offsets reported for Ga_2O_3 heterostructure and plotted in Fig. 5.27. The energy levels of the electron (E) and hole traps (H) are found above and below, respectively, the branch point. Within the error margins, this also applies to the hole trap H7.

Fig. 6.3 Experimental levels of irradiation-induced electron (blue) and hole traps (red) in 4*H*-, 6*H*-, and 3*C*-SiC polytypes

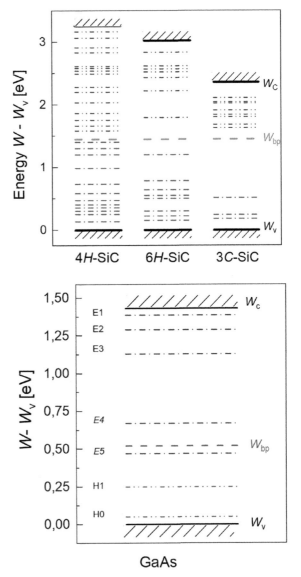

Fig. 6.4 Experimental levels of irradiation-induced electron and hole traps in GaAs from Pons & Bourgoini (1985) and Stievenard et al. (1990)

Fig. 6.5. Irradiation-induced
energy levels in the gap of
n-Ga_2O_3. Data compiled by
Kim et al. (2019). Branch-point
level from Fig. 5.27

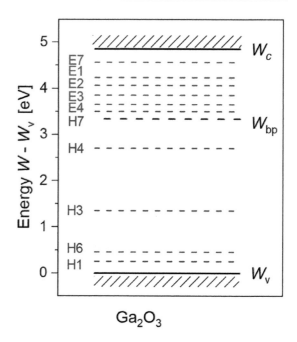

Ga$_2$O$_3$

Conclusions

The slope parameters S_X obtained in the Sects. 5.1 and 5.2 from the barrier heights of "*as ideal as possible*" and specifically laterally homogeneous *p*- and *n*-type *Schottky* contacts are plotted versus the optical dielectric susceptibility $\varepsilon_\infty - 1$ of the semiconductors in Fig. 7.1. Here, $A_X = 0.86\,\text{eV/Miedema}$ unit is used. Within the margins of error, there is no difference between the respective *p*- and *n*-type slope parameters. The dashed line is the linear least-squares fit

$$A_X/S_X - 1 = (0.07 \pm 0.04)(\varepsilon_\infty - 1)^{2.09 \pm 0.42} \tag{7.1}$$

to the data. This result excellently agrees with the theoretical prediction (3.53).

$$S_X \approx A_X/[1 + 0.08(\varepsilon_\infty - 1)^2]$$

On the other hand, it is somewhat lower than the correlation (1.9)

$$S_X = A_X/[1 + 0.10(\varepsilon_\infty - 1)^2],$$

which was obtained (Mönch 1986, 1987) from slope parameters in the determination of which the lateral homogeneity of the interfaces of the *Schottky* contacts was not yet explicitly considered (Schlüter 1978).

The additional data from LiF and BaF_2 (Pong & Paudyal 1981) and Xe (Jacob et al. 1987) *Schottky* contacts confirm that the IFIGS concept explains the physics of interfaces, regardless of whether the solid in question is traditionally considered a semiconductor, an ionic crystal or an insulator, or even has *van der Waals* bonds. The common and essential

© The Author(s), under exclusive license to Springer Nature Switzerland AG 2024   119
W. Mönch, *Electronic Structure of Semiconductor Interfaces*, Synthesis Lectures
on Engineering, Science, and Technology, https://doi.org/10.1007/978-3-031-59064-1_7

Fig. 7.1 Experimenmtal slope paraters S_X of p- and n-type *Schottky* contacts as a function of the optical dielcectic susceptibility $\varepsilon_\infty - 1$. The data are listed in Tables 5.1 and 5.2. The dashed line is the linear least-squares fit to the data without the Δ data. The dash-dotted line is taken from Fig. 1.9

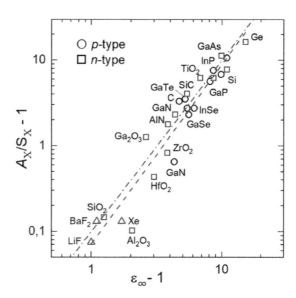

feature of all these solids is the presence of an energy gap separating the valence from the conduction bands.

In Fig. 7.2 the branch-point energies determined from the analysis of experimental data of p-type *Schottky* contacts as well as of heterostructures and summarized in Table 5.1 are plotted versus corresponding theoretical branch-point energies. The latter values have been calculated by Tersoff (1984) for Si and Ge, by Mönch (1996) for diamond, SiC, and the III-V compounds, and by Brudnyi et al. (2015) for the III-VI compounds. The linear least-squares fit

$$\Phi_{bp}^p(\exp) = (0.01 \pm 0.03) + (1.02 \pm 0.02)\Phi_{bp}^p(\text{theor}) \text{ [eV]} \qquad (7.2)$$

shows the excellent agreement between experimental and theoretical data. As the above analysis of the slope parameters, this finding also confirms Heine's (1965) conclusion that the interface-induced gap states, and hence the two terms of the IFIGS concept, the branch-point energies and the slope parameters, explain the band-structure alignment at all semiconductor interfaces.

The discussions in Sect. 3.5.2 of the formation of metal–semiconductor contacts from isolated metal adatoms to continuous metal overlayers on clean semiconductor surfaces and of adatom-induced shifts of the semiconductor core-levels revealed that the interface dipoles are excellently described by the partial ionic character of the covalent bonds between the metal and the semiconductor atoms directly at the interface. Together with

Fig. 7.2 Branch-point energies determined from experimental barrier heights of *p*-type *Schottky* contacts and valence-band offsets of heterostructures displayed in Figs. Figures 4.5 to 4.7 and 4.10 to 4.33, respectively, versus theoretical branch-point energies calculated by Tersoff (1984) for Si and Ge, by Mönch (1996) for diamond, SiC, GaN, GaP, GaAs, InN, and InP, and by Brudnyi et al. (2015) for the III-VI compounds

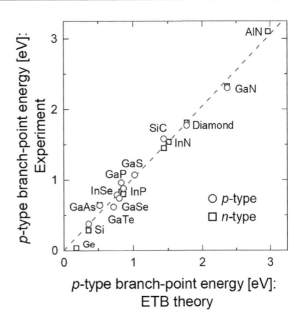

p-type branch-point energy [eV]: ETB theory

the above conclusion, this means that the partial ionic character of the covalent interface bonds and the interface-induced gap states are two equivalent explanations for the electronic properties of semiconductor interfaces.

It is often argued that the barrier heights of *Schottky* contacts and also of semiconductor heterostructures can be best investigated using *ab-initio* calculations. There are quite many of such efforts, in which individual metal–semiconductor combinations or heterostructures are considered and a variety of different approximations and assumptions are made. However, an overview of these approaches is far beyond the scope of this lecture. Here, results of three most recent investigations are briefly discussed. The first example is the *ab-initio* study of *Schottky* barrier heights of 10 different metal-diamond interfaces of Cheng et al. (2023). Their calculated barrier heights are plotted in Fig. 7.3 versus the electronegativity differences $X_m - X_C$ and not versus the metal work functions as done in their paper. The linear least-squares fit to these *ab-initio* calculated data

$$\Phi_{Bp} = (0.43 \pm 0.05) + (0.21 \pm 0.05)(X_m - X_C)[eV] \qquad (7.3)$$

has exactly the same slope as the prediction of the IFIGS concept but is shifted to lower energies by 1.34 eV compared to both the experimental data and the IFIGS prediction; see Sect. 5.1.1 and Fig. 7.1. This difference is almost as large as the shift of 1.4 eV caused by an H-interlayer in diamond *Schottky* contacts [Aoki & Kawarada, Kawarada et al. 1994] which is discussed in Sect. 2.5.2. This finding suggests that the *ab-initio* calculations of Cheng et al. either contain additional $Met^{+\Delta q} - C^{-\Delta q}$ interface dipoles

Fig. 7.3 Experimental (○)
and theoretical (□) *Schottky*
barrier heights from Fig. 5.1
and Cheng et al. (2023),
respectively. The IFIGS line is
drawn with $\Phi_{bp}^{p} = 1.77 \text{eV}$ and
$S_X = 0.21$ eV/Miedema-unit

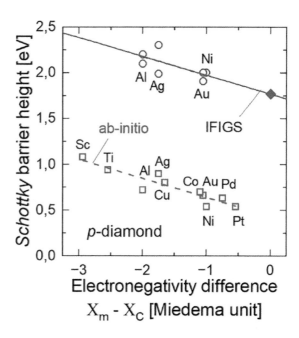

that are not present in real diamond *Schottky* contacts or give a branch point that is more
than 1 eV too close to the valence-band maximum.

The second example are the *ab-initio* calculations for Mg-, Ti-, Al-, Mn-, Ag-, Ni-,
and Pd/*p*-MgO *Schottky* contacts by Goniakowski and Noguera (2004). Almost simulta-
neously Lu et al. (2006b) prepared *Schottky* contacts on cleaved MgO(001) surfaces using
almost the same metals and determined their barrier heights using ultraviolet and X-ray
photoemission spectroscopy. The linear least-squares fits

$$\Phi_{Bp}^{hom} = (2.80 \pm 0.31) - (0.53 \pm 0.33)\left(X_m^{Mied} - X_{MgO}^{Mied}\right)[\text{eV}] \qquad (7.4)$$

to the calculated data and

$$\Phi_{Bp}^{hom} = (4.24 \pm 0.26) - (0.59 \pm 0.27)\left(X_m^{Mied} - X_{MgO}^{Mied}\right)[\text{eV}] \qquad (7.5)$$

to the experimental results give the same slope parameters but the branch-point energies
differ by 1.44 eV.[1] This experimental slope parameter agrees with the value 0.65 eV/
Miedema-unit predicted by relation (7.1) for MgO interfaces. The experimental branch-
point energy $\Phi_{bp}^{p} = 4.24$eV is, within the margins of error, consistent with the value
of 3.92 eV obtained from an analysis of experimental valence-band offsets of MgO

[1] It should be explicitly mentioned that both Goniakowski and Noguera (2004) and Lu et al. (2006b)
plotted their *Schottky* barrier heights versus Pauling's electronegativities rather than the work functions of the metals used.

heterostructures by Mönch (2011). Thus, also in this second example, the *ab-initio* calculations of Goniakowski and Noguera (2004) either contain additional $Met^{+\Delta q} - MgO^{-\Delta q}$ interface dipoles that are not present in real MgO *Schottky* contacts or result in a branch point that is more than 1 eV too close to the top of the valence band.

The third example are the *ab-initio* calculations for Pt/n-Ga$_2$O$_3$ *Schottky* contacts of Therrrien et al. (2021). They obtained a barrier height of 1.9 eV. Their result is by approximately 0.6 eV larger than both the average of the experimentally observed values and the prediction by the IFIGS concept; see Fig. 5.25. This finding again suggests that additional $Pt^{+\Delta q} - Ga_2O_3^{-\Delta q}$ dipoles are present in the *ab-initio* approach used, but not in real Pt/Ga$_2$O$_3$ contacts, or that the *ab-initio* calculations place the branch-point by approximately 0.6 eV too close to the valence-band top of Ga$_2$O$_3$.

Walukiewicz (2001) claimed that the branch points of the IFIGS are generally located by 4.9 eV below the vacuum level W_{vac}. It would indeed be an attractive idea if the branch point energies could be calculated so easily. Figure 7.4 displays the branch-point work-functions $W_{vac} - W_{bp} = I - \Phi_{bp}^p$ as a function of the electronegativities of the semiconductors which are selected only as arrangement parameters. Here, experimental ionization energies[2] $I = W_{vac} - W_v$ and the experimental branch-point energies summarized in Tables 5.1 and 5.2 are considered only. The average value of the experimental branch-point work-functions, 4.7 ± 0.7 eV, is close to the postulated value of 4.9 eV and the same also holds for the data of Si, Ge, a few of the III-V and II-VI compounds, InGaZnO$_4$ and LaAlO$_3$. However, the majority of the data deviate by more than ± 0.2 eV from the general value postulated by Walukiewicz. This finding is plausible because the branch points are part of the bulk band structure, but not surface properties, so the vacuum level is not a relevant reference.

To summarize, the alignment of the band structures at semiconductor interfaces, i.e., the barrier heights of *Schottky* contacts and the band offsets at semiconductor heterostructures, are determined by a continuum of interface-induced gap states (IFIGS), as first proposed by Heine (1965). These gap states originate from the complex band structure of the semiconductors and can be described by two bulk parameters. One is the branch point where the charge character of the induced gap states changes from acceptor-like to donor-like, i.e., it has the character of a charge-neutrality level. This energy level is located slightly below the middle of the energy gap at the mean-value k-point of the semiconductor. The second parameter describes the interface dipole. It can be explained in two ways. From a physical point of view, it is given by the density of states and the decay length of the interface-induced gap states at their branch point. Both quantities require extensive calculations. From the chemist's point of view, Pauling's concept (1939/1960) of the partial ionic character of covalent bonds in molecules can be also

---

[2] **Figure 7.4**. Experimental ionization energies of semiconductors from Bersch et al. (2008), Kumar et al. (2012), Lee et al. (2012), Liu et al. (2011, 2012), Mönch (2001), Mohamed et al. (2006), Nazarzadehmoaf et al. (2014), Schafranek & Klein (2006), Thomas et al. (1980), Wu & Kahn (2000), Xiong et al. (2007).

Fig. 7.4 Branch-point work functions $W_{vac} - W_{bp} = I - \Phi^p_{bp}$ as a function of the electronegativities of the semiconductors. The dashed line indicates Walukiewicz's (2001) general postulate $W_{vac} - W_{bp} = 4.9 eV$. From Mönch (2018)

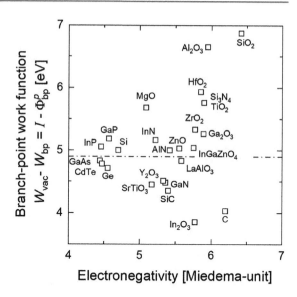

transferred to the chemical bonds at semiconductor interfaces. The interface dipoles at semiconductor interfaces then vary proportional to the difference of the electronegativities of the metal and the semiconductor or of the two semiconductors in contact. The corresponding proportionally or slope parameters were found to depend quadratically on the optical susceptibility $\varepsilon_\infty - 1$ of the semiconductor (Mönch 1985). A *"bridge"* between the physicists' approach and that of the chemists is the linear correlation found between the core-level shifts induced by metal as well as non-metal adatoms on semiconductor surfaces and the differences between their electronegativities. Finally, the barrier heights of *Schottky* contacts and the band offsets of heterostructures are consistently and quantitatively explained by the IFIGS-and-electronegativity concept. The same conclusion also applies to adatom-induced surface or gap states and the saturation *Fermi*-level positions due to irradiation-induced native defects in semiconductors.

Appendix

© The Editor(s) (if applicable) and The Author(s), under exclusive license
to Springer Nature Switzerland AG 2024
W. Mönch, *Electronic Structure of Semiconductor Interfaces*, Synthesis Lectures
on Engineering, Science, and Technology, https://doi.org/10.1007/978-3-031-59064-1

Table A.1 Atomic electronegativities after Pauling (1939/1960) and Miedema et al. (1973, 1980). Miedema values in brackets are obtained using

$$X_{Mied} = 1.93\,X_{Paul} + 0.87 \;(\text{Mönch 1993})$$

1	2	3	4	5	6	7	8	9	10	11	12	13	14	15	16	17	18
H 1 2.20																	He 2
Li 3 0.98 2.85	Be 4 1.57 4.20											B 5 2.04 4.75	C 6 2.55 6.20	N 7 3.04 7.00	O 8 3.44 (7.50)	F 9 3.98	Ne 10
Na 11 0.93 2.70	Mg 12 1.31 3.45											Al 13 1.61 4.20	Si 14 1.90 4.70	P 15 2.19 5.10	S 16 2.58 (5.84)	Cl 17 3.16	Ar 18
K 19 0.82 2.25	Ca 20 1.00 2.55	Sc 21 1.36 3.25	Ti 22 1.54 3.65	V 23 1.63 4.25	Cr 24 1.66 4.65	Mn 25 1.55 4.45	Fe 26 1.83 4.93	Co 27 1.88 5.10	Ni 28 1.91 5.20	Cu 29 1.90 4.55	Zn 30 1.65 4.10	Ga 31 1.81 4.10	Ge 32 2.01 4.55	As 33 2.18 4.80	Se 34 2.55 (5.79)	Br 35 2.96	Kr 36
Rb 37 0.82 2.10	Sr 38 0.95 2.40	Y 39 1.22 3.20	Zr 40 1.33 3.40	Nb 41 1.60 4.00	Mo 42 2.16 465	Tc 43 1.90 5.30	Ru 44 2.20 5.40	Rh 45 2.28 5.40	Pd 46 2.20 5.45	Ag 47 1.93 4.45	Cd 48 1.69 4.05	In 49 1.78 3.90	Sn 50 1.96 4.15	Sb 51 2.05 4.40	Te 52 2.10 (4.99)	I 53 2.66	Xe 54
Cs 55 0.79 1.95	Ba 56 0.89 2.32	La 57 1.10 3.05	Hf 72 1.30 3.55	Ta 73 1.50 4.05	W 74 2.36 4.80	Re 75 1.90 5.40	Os 76 2.20 5.40	Ir 77 2.20 5.55	Pt 78 2.28 5.65	Au 79 2.54 5.15	Hg 80 2.00 4.20	Tl 81 2.04 3.90	Pb 82 2.33 4.10	Bi 83 2.02 4.15	Po 84 2.00	At 85 2.20	Rn 86
Fr 87 0.7	Ra 88 0.90	Ac 89 1.10															

Ce 58 1.12	Pr 59 1.13	Nd 60 1.14	Pm 61 1.13	Sm 62 1.17	Eu 63 1.20	Gd 64 1.20	Tb 65 1.20	Dy 66 1.22	Ho 67 1.23	Er 68 1.24	Tm 69 1.25	Yb 70 1.10	Lu 71 1.27
Th 90 1.30 3.30	Pa 91 1.50	U 92 1.38 4.05											

References

Aboelfotoh, M. O., Fröjdh, C., & Petersson, C. S. (2003). Phys. Rev. B. 67, 075312. Figs. 5.2, 5.15.

Acar, S., Karadeniz, S., Tugluoglu, N., Selçuk, A. B, & Kasap, M. (2004). Appl. Surf. Sci. 233, 373. Sect. 2.4.

Adams, W. G., & Day, R. E. (1876). Proc. Roy. Soc. 25, 113. Sect. 1.1.

Afanas'ev, V. V., & Stesmans, A. (2000). Appl. Phys. Lett. 77, 2024. Figs. 5.16, 5.31, 5.32.

Afanas'ev, V. V., Stesmans, A. (2004). Appl. Phys. Lett. 84, 2319. Fig. 5.13.

Afanas'ev, V. V., Bassler, M., Pensl, G., Schulz, M. J., & Stein von Kamienski, E. (1996). J. Appl. Phys. 79, 3108. Figs. 5.12, 5.16, 5.31, 5.32.

Afanas'ev, V. V., Houssa, M., Stesmans, A., & Heyns, M. M. (2001). Appl. Phys. Lett. 78, 3073. Figs. 5.12, 5.29, 5.31, 5.32.

Afanas'ev, V. V., Stesmans, A., Chen, F., Shi, X., & Campbell, S. A. (2002a). Appl. Phys. Lett. 81, 1053. Fig. 5.12.

Afanas'ev, V. V., Houssa, M., Stesmans, A., & Heyns, M. M. (2002b). Appl. Phys. 91, 3079. Fig. 5.28.

Afanas'ev, V. V., Houssa, M., Stesmans, A., Adriaenssens, G. J., & Heyns, M. M. (2002c). J. Non-Cryst. Solids 303, 69. Fig. 5.30.

Afanas'ev, V. V., Chou, H. -Y., Houssa, M., Stesmans, A., Lamagna, L., Lamperti, A., Molle, A., Vincent, B., & Brammertz, G. (2011). Appl. Phys. Lett. 99, 172101. Fig. 5.13.

Ahn, S., Ren, F., Yuan, L., Pearton, S. J., & Kuramata, A. (2017). ECS J. Solid State Sci. Technol. 6, P68. Fig. 5.25.

Akklilic, K., Aydin, M.E., & Türüt, A. (2004). Physica Scripta. 70, 364 Fig. 3.11.

Alay, J. L., & Hirose, M. (1997). J. Appl. Phys. 81, 1606. Figs. 5.12, 5.31. 5.32.

Alferov, Zh. I. (2001). Rev. Mod. Phys. 73, 767. Sect. 1.3.

Alferov, Zh. I., & Kazarinov, R. F. (1963). Russian Inventor's Certificate No. 181737. Sect. 1.3.

Alfieri, G., & Kimoto, T. (2009). Mater. Sci. Forum 615–617, 389. Fig. 6.3.

Alfieri, G., & Kimoto, T. (2011). J. Phys. Condens. Matter 23, 065803. Fig. 6.3.

Alfieri, G., Mihail, A., Ayed, H. M., Svensson, B. G., Hazdra, P., Godignon, P., Millan, J., & Kicin, S. (2016). Mater. Sci. Forum 858, 308. Fig. 6.3.

Alivov, Y., Bo, X., Fan Qian, F., Johnstone, D., Litton, C., Morkoç, H. (2006). MRS Online Proceedings 957, 1021. Fig. 5.16.

Alivov, Y. I., Xiao, B., Akarca-Biyikli, S., Fan, Q., Morkoç, H., Johnstone, D., Lopatiuk-Tirpak, O., Chernyak, L., & Litton, W. (2008). J. Phys.: Condens. Matter 20, 085201. Fig. 5.20.

Allen, F. G., & Gobeli, G. W. (1966). Phys. Rev. 144, 558. Sect. 3.5.1.

Allen, N., Ciarkowski, T., Carlson, E., Chakraborty, A., Guido, L. (2020) AIP Advances 10, 015116. Fig. 5.19.

© The Editor(s) (if applicable) and The Author(s), under exclusive license to Springer Nature Switzerland AG 2024

W. Mönch, *Electronic Structure of Semiconductor Interfaces*, Synthesis Lectures on Engineering, Science, and Technology, https://doi.org/10.1007/978-3-031-59064-1

Almeida, J., dell'Orto, T., Coluzza, C., Margaritondo, G., Bergossi, O., Spajer, M., & Courjon, D. (1996). Appl. Phys. Lett. 69, 2361. Fig. 5.4.

Altuntas, H., Donmez, I., Ozgit-Akgun, C., & Biyikli, N. (2014). J. Alloys Comp. 593, 190. Fig. 5.25.

Amy, F., Wan, A. S., Khan, A., Walker, J. F., & McKee, R. A. (2004). J. Appl. Phys. 96, 1635. Figs. 5.12, 5.34.

Anderson, R. L. (1962). Solid State Electron. 5, 341. Sects. 1.3, 3.8.1.

Andres, S., Pettenkofer, C., Speck, F., & Seylle, T. (2008). J. Appl. Phys. 103, 103720. Fig. 5.16.

Aniltürk, Ö. S., & Turan, R. (1999). Semicond. Sci. Technol. 14, 1060. Fig. 2.21.

Aoki, M., & Kawarada, H. (1994). Jpn. J. Appl. Phys. 33, L708. Sects. 2.5.2, 5.3. Chapt. 7. Fig. 2.18.

Aoyama, K., Ueno, K., Kobayashi, A., & Fujioka, H. (2022). Appl. Phys. Lett. 121, 232103. Fig. 5.3.

Arulkumaran, S., Arokiaraj, J., Dharmarasu, N., & Kumar, J. (1996). Nucl. Instr. and Meth. B 116, 519. Fig. 5.17.

Asimov, A., Ahmetoglu, M., Kucur, B., Özer, M., & Güzel, T. (2013). Optoelectr. Adv. Mater. - Rapid Commun. 7, 490. Fig. 5.5.

Aydin, M. E., Akkilic, K., & Kiliçoğlu, T. (2004). Appl. Surf. Sci. 225, 318. Fig. 2.18.

Aydin, M. E., Yıldırım, N., & Türüt, A. (2007). J. Appl. Phys. 102, 043701. Fig. 5.15.

Baldereschi, A. (1972). Bull. Am. Phys. Soc. 17, 237. Sects. 4.1, 4.2.

Baldereschi, A. (1973). Phys. Rev. B 7, 5212. Sects. 1.2, 4.1, 4.2.

Ban, D. -Y., Fang, R. -C., Xue, J. -G., Lu, E. -D., & Xu, P. -S. (1997). Chin. Phys. Lett. 14, 609. Fig. 5.13.

Ban, D. -Y., Xue, J. -G., Fang, R. -C., Xu, S. -H., Lu, E. -D., Xu, P. -S. (1998). J. Vac. Sci. Technol. B 16, 989. Figs. 5.31, 5.32.

Baraff, G. A., & Schlüter, M. (1984). Phys. Rev. B 30, 3460. Sect. 4.1.

Bardeen, J. (1947). Phys. Rev. 71, 717. Sects. 1.2. 3.2.

Baumeier, B., Krüger, P., & Pollmann, J. (2007). Phys. Rev. B 76, 205404, and private communications 2009. Sect. 4.1. Fig. 4.2.

Bechstedt, F., Fuchs, F., & Kresse, G. (2009). Phys. Status Solidi B 246, 1877. Sect 4.1.

Belabbes, A., de Carvalho, L. C., Schleife, A., & Bechstedt, F. (2011). Phys. Rev. B 84, 125108. Sects. 1.2, 4.1, 4.2. Fig. 4.5.

Belabbes, A., Panse, C., Furthmüller, J., & Bechstedt, F. (2012). Phys. Rev. B 86, 075208. Sects. 1.2, 4.1, 4.2. Fig. 4.5.

Belabbes, A., Botti, S., & Bechstedt, F. (2022). Phys. Rev. B 106, 085303. Sect. 1.2.

Bell, L. D., & Kaiser, W. J. (1988). Phys. Rev. Lett. 61, 2368. Sect. 2.4.

Bell, L. D., Smith, R. P., McDermott, B. T., Gertner, E. R., Pittman, R., Pierson, R. L., & Sullivan, G. J. (1998). Appl. Phys. Lett. 72, 1590. Fig. 5.19.

Bersch, E., Rangan, S., Bartynski, R. A., E. Garfunkel, E., & Vescovo, E. (2008). Phys. Rev. B 78, 085114. Fig. 5.12.

Bersch, E., Rangan, S., & Bartynski, R. A. (2008). Phys. Rev. B 78, 085114. Fig. 7.4.

Berthod, C., Binggeli, N., & Baldereschi, A. (1996). Europhys. Lett. 36, 67. Tab. 4.2.

Bethe, H. A. (1942). MIT Radiation Lab. Rep. 43–12. Sect. 2.1.

Bhat, T. N., Kumar, M., Rajpalke, M. K., Roul, B., Krupanidhi, S. B., & Sinha, N. (2011). J. Appl. Phys. 109, 123707. Fig. 5.12.

Bhuiyan, A.S., Martinez, A., Esteve, D. (1988) Thin. Solid Films. 161, 93. Fig. 5.17.

Bhuiyan, A. F. A. U., Feng, Z., Huang, H. -L., Meng, L., Hwang, J., & Zhao, H. (2021). APL Mater. 9, 101109. Fig. 5.29,

Binari, S. C., Dietrich, B., Kelner, G., Rowland, L. B., Doverspike, K., & Gaskill, D. K. (1994). Electron. Lett. 30, 909. Fig. 5.19.

Björk, M. T., Schmid, H., Besssire, C. D., Moselund, K. E., Ghoneim, H., Karg, S., Lörtscher, E., & Riel, H. (2010). Appl. Phys. Lett. 97, 163501. Fig. 5.12.

Blauärmel, A., Brauer, M., Hoffmann, V., & Schmidt, M. (2000). Appl. Surf. Sci. 166, 108. Figs. 3.11, 5.11.

Bohlke, W. H., Kalma, A. H., & Corelli, J. C. (1966). J. Appl. Phys. 39, 5498. Figs. 6.1, 6.2.

Boiko, V. M., Brudnyi, V. N., Ermakov, V. S., Kolin, N. G., & Korulin, A. V. (2015). Semiconductors 49, 763. Figs. 6.1, 6.2.

Bose, J. C. (1904). U. S. Patent No. 775,840. Sects. 1.1, 3.1.

Bose, D. N., & Pal, S. (1997). Phil. Mag. B 75, 211. Fig. 5.6.

Botzakaki, M.A., Skoulatakisb, G., Papageorgiouc, G.P., Krontiras, C.A. (2020). Semicon-ductors, 54, 543. Fig. 5.13.

Brattain, W. H. (1940). see Bardeen, J. (1947). Sect. 3.1.

Braun, F. (1874). Poggendorfs Ann. Physik und Chemie 153, 556. Sect. 1.1. For an English translation, see: *Semiconductor Devices: Pioneering Papers*, edited by S. M. Sze, p. 377 (World Scientific, Singapore and Teaneck, NJ, 1991).

Braun, F. (1906). Elektrotechn. Zeitschr. 52, 1199. Sect. 1.1.

Brudnyi, V. N., & Grinyev, S. N. (1998). Semiconductors 32, 284. Figs. 6.1, 6.2.

Brudnyi, V. N., & Kosobutsky, A. V. (2017). Superlattices and Microstructures 111, 499. Sect. 5.1.1. Fig. 4.7.

Brudnyi, V. N., Krivov, M. A., & Potapov, A.I. (1982). Sov. Phys. 25, 37. Figs. 6.1, 6.2.

Brudnyi, V. N., Grinyaev, S. N., & Stepanov, V. E. (1995). Physica B 212, 429. Sects. 1.2, 4.1, 5.2.3. Fig 4.7.

Brudnyi, V. N., Kolin, N. G., & Potapov, A. I. (2003). Semiconductors 37, 39. Figs. 6.1, 6.2.

Brudny, V. N., Kolin, N. G., Merkurisov, D. I., & Novikov, V. A. (2005). Semiconductors 39, 499. Figs. 6.1, 6.2.

Brudnyi, V. N., Kosobutsky, A. V., & Kolin, N. G. (2008). Russian Phys. J. 51, 1270. Sect. 4.1. Fig. 4.7.

Brudnyi, V. N., Verevkin, S. S., Kolin, N. G., & Korulin, A. V (2012). Russ. Phys. J. 55, 53. Figs. 6.1, 6.2.

Brudnyi, V. N., Sarkisov, S. Yu, & Kosobutsky, A. V. (2015). Semicond. Sci. Technol. 30, 115019. Sect. 5.1.1, 5.1.2, 5.2. Fig. 5.7, 5.8.

Buchanan, D. A., McFeely, F. R., & Yurkas, J. J. (1998). Appl. Phys. Lett. 73, 1676. Fig. 5.30.

Buzio, R., Gerbi, A., He, Q., Qin, Y., Mu, W., Jia, Z., Tao, X., Xu, G., & Long, S. (2020). Adv. Electron. Mater. 1901151. Fig. 5.25.

Cañas, J., Alba, G., Leinen, D., Lloret, F., Gutierrez, M., Eon, D., Pernot, J., Gheeraert, E., & Araujo, D. (2021). Appl. Surf. Sci. 535, 146301. Fig. 5.10.

Cantoni, M, Petti, D., Rinaldi, C., & Bertacco, R. (2011). Appl. Phys. Lett. 98, 032104. Fig. 5.13.

Cardona, M., & Christensen, N. E. (1987). Phys. Rev. B 35, 6182. Sect. 4.1.

Carey IV, P. H., Ren, F., Hays, D. C., Gila, B. P, Pearton, S. J., Jang,S., & Kuramata, A. (2017). Appl. Surf. Sci. 422, 179. Figs. 5.26, 5.27.

Cetin, H., Sahin, B., Ayyildiz, E., & Türüt, A. (2004). Semicond. Sci. Technol. 19, 1113. Figs. 2.18, 3.11.

Chadi, D. J., & Cohen, M. L. (1973). Phys. Rev. B 8, 5747. Sect. 4.1.

Chambers, S. A. Chambersa, Y. Liang Z. Yu, Droopad, R., & J. Ramdani, J. (2001). J. Vac. Sci. Technol. A 19, 934. Fig. 5.34.

Chambers, S. A., Droubay, T., Kasper,T. C., & Gutowski, M. (2004). J. Vac. Sci. Technol. B 22, 2205. Figs. 5.12, 5.34.

Chambers, S. A., Ohsawa, T., Wang, C. M., Lyubinetsky, I., & Jaffe, J. E. (2009). Surf. Sci. 603 771. Fig. 5.34.

Chambers, S. A., Du Y., Comes, R. B., Spurgeon, S. R., & Sushko, P. V. (2017a). Appl. Phys. Lett. 110, 082104. Fig. 5.13.

Chambers, S. A., Schütz, P., Pfaff, F., Scheiderer, P., Chen, Y. Z., Pryds, N., Gorgoi, M., Sing, M., & Claessen, R. (2017b). Appl. Phys. Lett. 110, 082104. Fig. 5.34.

Chang, S. -H., Chen, Z. -Z., Huang, W., Liu, X. -C., Chen, Bo-Y., Li, Z. -Z., & Shi Er-W. (2011). Chin. Phys. B 20, 116101. Fig. 5.16.

Chang, S. -H., Chen, Z. -Z., Huang, W., Liu, X. -C., Chen, B. -Y., Li, Z. -Z., & Shi, Er-W. (2011). Chin. Phys. B 20, 116101. Figs. 5.26, 5.27.

Chao, Y. -C., Johansson, L. S. O., & Uhrberg, R.I. G. (1996). Phys. Rev. B 54, 5901. Sect. 3.5.2.

Chaoudhary, S., Dewasi, A., Ghosh, S., Choudhary, R. J., Phase, D. M., Ganguli, T., Rastogi, V., Pereira, R. N., Sinopoli, A., Aïssa, B., & Mitra, A. (2022). Thin Solid Films. 743, 139077. Fig. 5.12.

Chen, T. P., Lee, T. C., Ling, C. C., Beling, C. D., & Fung, S. (1993a). Solid-State Electron. 36, 949. Sect. 2.6.1. Figs. 3.11, 5.11.

Chen, T. P., Lee, T. C., Ling, C.C., Beling, C.D., & Fung, S. (1993b) Solid-State Electronics. 36, 949.

Chen, C. -H., Aballe, L., Klauser, R., Kampen, T. U., & Horn, K. (2005a). J. Electron Spectroscop. Relat. Phenom. 144, 425. Fig. 5.20.

Chen, H., Aballe, L., Klauser, R., Kampen, T. U., & Horn, K. (2005b). J. Electron Spectrosc. Related Phenomena 144–147, 425C. Fig. 5.16.

Chen, J. -J., Gila, B. P., Hlad, M., Gerger, A., Ren, F., Abernathy, C. R., & Pearton, S. J. (2006). Appl. Phys. Lett. 88, 042113. Fig. 5.20.

Chen, Q., Feng, Y. P., Chai, J. W., Zhang, Z., Pan, J. S., & Wang S. J. (2008). Appl. Phys. Lett. 93, 052104. Fig. 5.16.

Chen, J., Lv, J., & Wang, Q. (2016a). Thin Solid Films 616, 145. Fig. 3.12.

Chen, Z., Nishihagi, K., Wang, Xu, Saito, K., Tanaka, T., Nishio, M., Arita, M., & Guo, Q. (2016). Appl. Phys. Lett. 109, 102106. Fig. 5.12.

Chen, J. -X., Tao, J. -J., Ma, H. -P., Zhang, H., Feng, Ji-J., Liu, W. -J., Xia, C., Lu, H. -L., & Zhang, D. W. (2018). Appl. Phys. Lett. 112, 261602. Figs. 5.22, 5.23.

Chen, J., Tao, R., Wang, G., Yin, Z., Zeng, L., Yu, X., Zhang, S., Wu, Y., Lic, Z., & Zhang, X. (2023). J. Mater. Chem. C 11, 5324. Fig. 5.10.

Chen, Z., Nishihagi, K., Wang, Xu, Saito, K., Tanaka, T., Nishio, M., Arita, M., & Guo, Q. (2016). Appl. Phys. Lett. 109, 102106. Figs. 5.26, 5.27.

Chen, J. -X., Tao, J. -J., Ma, H. -P., Zhang, H., Feng, Ji-J., Liu, W. -J., Xia, C., Lu, H. -L., & Zhang, D. W. (2018). Appl. Phys. Lett. 112, 261602. Figs. 5.26, 5.27.

Chen, X. H., Chen, Y. T., Ren, F. -F., Gu, S. L., Tan, H. H., Jagadish, C., & Ye, J. D. (2019). Appl. Phys. Lett. 115, 202101. Figs. 5.26, 5.27.

Chen, Y., Ning, H., Kuang, Y., Yu, X. -X., Gong, H. -H., Chen, X., Ren, F. -F., Gu, S., Zhang, R., Zheng, Y., Wang, X., & Ye, J. (2022). Sci. China-Phys. Mech. Astron. 65, 277311. Figs. 5.26, 5.27.

Cheng, C. -P., Chen, W. -S., Lin. K. -Y., Wei, G. -J., Cheng, Yi-T., Lin, Y. -H., Wan, H. -W., Pi, T. -W., Tung, R. T., Kwo, J., & Hong, M. (2017). Appl. Surf. Sci. 393, 294. Fig. 5.5.

Cheng, C., Zhang, Z., Sun, X., Gui, Q., Wu, G., Dong, F., Zhang, D., Guo, Y., & Liu, S. (2023). Appl. Surf. Sci. 615, 156329. Sect. 5.3. Chapt. 7. Fig. 5.37, 7.3.

Chiam, S. Y., Chim, W. K., Pi, C., Huan, A. C. H., Wang, S. J., Pan, J. S., Turner, S., & Zhang, J. (2008). J. Appl. Phys. 103, 083702. Fig. 5.12.

Chin, V. W. L., Green, M. A., & Storey, J. W. V. (1993). Solid-State Electron. 36, 1107, Fig. 2.20.

Choi, J., Puthenkovilakam, R., & Chang, J. P. (2005). Appl. Phys. Lett. 86, 192101. Figs. 5.16, 5.22, 5.23.

Chou, M. Y., Cohen, M. L., & Louie, S. G. (1985). Phys. Rev. B 32, 7979. Sect. 2.5.1, Fig. 2.16.

Chui, C. O., Lee, D. -I., Singh, A. A., Pianetta, P. A., & Saraswat, K. C. (2005). J. Appl. Phys. 97, 113518. Fig. 5.13.

Clavel, M. B., & Hudait, M. K. (2017). IEEE Electron Device Lett. 38, 1196. Fig. 5.13.

Cleland, J. W., & Crawford, Jr., J. H. (1955). Phys. Rev. 100, 1614. Figs. 6.1, 6.2.

Cohen, J., Vilms, J., & Archer, R. J. (1969). Hewlett-Packard R&D Report AFCRL-69–0287. Figs. 3.11, 5.11.

Cola, A., Lupo, M. G., Vasanelli, L., & Valentini, A. (1992). J. Appl. Phys. 71, 4966. Fig. 5.17.

Connelly, D., Faulkner, C., Grupp, D. E., & Harris, J. S. (2004). IEEE Trans. Nanotechnol. 4. 98. Fig. 2.18.

Connelly, D., Faulkner, C., Clifton, P. A., & Grupp, D. E. (2006). Appl. Phys. Lett. 88, 0121051. Fig. 2.18.

Constantinidis, G., Kuzmic, J., Michelakis, K., & Tsagarak, K. (1998). Solid-State Electron. 42, 253. Fig. 5.15.

Cook jr., T. E., Fulton, C. C., Mecouch, W. J., Tracy, K. M., Davis, R. F., Hurt, E. H., Lucovsky, G., & Nemanich, R. J. (2003a). J. Appl. Phys. 93, 3995. Figs. 5.20, 5.31, 5.32.

Cook, jr., T. E., Fulton, C. C., Mecouch, W. J., & Davis, R. F. (2003b). J. Appl. Phys. 94, 3949. Fig. 5.20.

Cook, jr., T. E., Fulton, C. C., Mecouch, W. J., Davis, R. F., Lucovsky, G., & Nemanich, R. J. (2003c). J. Appl. Phys. 94, 7155. Figs. 5.20, 533.

Coskun, C., Biber, M., & Efeoglu, H. (2003). Appl. Surf. Sci. 211, 360. Fig. 5.6.

Cowley, A. M., & Sze, S. M. (1965). J. Appl. Phys. 36, 3212. Sects. 1.1, 1.2, 3.2, 3.6, 3.7, 4.2.

Craft, H. S., Collazo, R., Losego, M. D., Mita, S., Sitar, Z., & Maria, J. -P. (2007). J. Appl. Phys. 102, 074104. Fig. 5.20.

Crowell, C. R., Spitzer, W. G., Howarth, L. E., & LaBate, E. E. (1962). Phys. Rev. 127, 2006. Figs. 3.11, 5.11.

Crowell, C. R., Sze, S. M., & Spitzer, W. G. (1964). Appl. Phys. Lett. 4, 91. Figs. 3.11, 5.11.

Dalapati, G. K., Oh, H. -J., Lee, S. J., Sridhara, A., Wong, A. S. W., & Chi, D. (2008) Appl. Phys. Lett. 92, 042120. Fig. 5.18.

Dalibor, T., Pensl, G., Matsunami, H., Kimoto, T., Choyke, W. J., Schöner, A., & Nordell, N. (1997) Physica status solidi (a). 162, 199. Fig. 6.3.

Danno, K., & Kimoto, T. (2006). J. Appl. Phys. 100, 113728. Fig. 6.3.

Danno, K., & Kimoto, T. (2007a). Mater. Sci. Forum 556–557, 331. Fig. 6.3.

Danno, K., & Kimoto, T. (2007b). J. Appl. Phys. 101, 103704. Fig. 6.3.

Das, P., Jones, L. A. H., Gibbon, J. T., Dhanak, V. R., Partida-Manzanera, T., Roberts, J. W., Potter, P., Chalker, P. R., Cho, S. -J., Thayne, I. G., Mahapatra, R., & Mitrovic, I. Z. (2020). ECS J. Solid State Sci. Technol. 9, 063003. Fig. 5.20.

Davydov, D. V., Lebedev, A. A., Kozlovski, V. V., Savkina, N. S., & Strel'chuk, A. M. (2001). Physica B 308–310, 641. Fig. 6.3.

Deal, B. E., Snow, E. H., & Mead, C. A. (1966). J. Phys. Chem Solids 27, 1873. Fig. 5.30

Defives, D., Noblanc, O., Dua, C., Brylinski, C., Barthula, M., & Meyer, F. (1999). Mater. Sci. Engin. B 61–62, 395. Fig. 5.15.

Demkov, A. A., Fonseca, L. R. C., Verret, E., Tomfohr, J., & Sankey, O. F. (2005). Phys. Rev. B 71, 195306. Tab. 4.2.

Deng, Y., Yang, Z., Xu, T., Jiang, H., Ng, K. W. Liao. C., Su, D., Pei, Y., Chen, Z., Wang, G., & Lu, X. (2023). Appl. Surf. Sci. 622, 156917. Figs. 5.26, 5.27.

Deniz, A. R., Çaldıran, Z., Metin, Ö., Can, H., Meral., K., & Aydoğan, S. (2014). J. Mater. Sci. Semicond. Processing 27, 163. Fig. 5.5.

Dharmarasu, N., Arulkumaran, S., Sumathi, R. R., Jayavel, P., Kumar, J., Magudapathy, P., & Nair, K. G. M. (1998). Nucl. Instr. and Meth. B 140, 119. Fig. 5.17.

DiDio, M., Cola, A., Lupo, M. G., & Vasanelli, L. (1995). Solid-State Electron. 38, 1923. Fig. 5.17.

DiMaria, D. J. (1974). J. Appl. Phys. 45, 5454. Fig. 5.28.

DiMaria, D. J., & Arnett, P. C. (1975). Appl. Phys. Lett. 26, 711. Fig. 5.33.

Dimoulas, A., Tsipas, P., Sotiropoulos, A., & Evangelou, E. K. (2006). Appl. Phys. Lett. 89, 252110. Fig. 5.13.

Ding, S. A., Barman, S. R., Horn, K., Yang, H., Yang, B., Brandt, O., & Ploog, K. (1997). Appl. Phys. Lett. 70, 2407. Figs. 5.18, 5.20.

Dixit, V. K., Kumar, S., Singh, S. D., Khamari, S. K., Kumar, R., Tiwari, P., Phase, D. M., Sharma, T. K., & Oak, S. M. (2014). Appl. Phys. Lett. 104, 092101. Fig. 5.13.

Doğan, H., Korkut, H., Yıldırım, N., & Türüt, A. (2007). Appl. Surf. Sci. 253, 7467. Fig. 5.17

Doğan, H., & Elagoz, S. (2014) Physica E. 63, 186. Fig. 5.19.

Dong, Y. F., Mi, Y. Y., Feng, Y. P., Huan, A. C., & Wang, S. J. (2006a). Appl. Phys. Lett. 89, 122115. Tab. 4.2.

Dong, Y.F., Wang, S.J., Feng, Y.P., Huan, A.C.H. (2006b). Physical Review B. 73, 045302.

Donnelly, J. P., & Hurwitz, C. E. (1977). Solid-State Electron. 20, 727. Figs. 6.1, 6.2.

Du, Y., Sushko, P. V., Spurgeon, S. R., Bowden, M. E., Ablett, J. M., Lee, T. -L., Quackenbush, N. F., Woicik J. C., & Chambers, S. A. (2018a). Phys. Rev. Mater. 2, 094602. Fig. 5.13.

Du, J. -J., Xu, S. R., Zhang, J. -C., Li, P. -X., Lin, Z. -Yu, Zhao, Y., Peng, R. -S., Fan, X. -M., Tao, H. -C., & Hao, Y. (2018b). Materials Letters 230, 135. Figs. 5.22, 5.23.

Duan, T. L., Pan, J. S., & Ang, D. S. (2013). Appl. Phys. Lett. 102, 201604. Fig. 5.20.

Duman, S., Kaya, F. S., Dogan H., & Turgut, G. (2022). Radiation Phys. Chem. 193, 109992. Fig. 5.5.

Dunwoody, H. M. C. (1906). US Patent 837,616. Sect. 1.1.

Dyment, J. C., North, J. C., & D'Asaro, L. A. (1973). J. Appl. Phys. 44, 207. Figs. 6.1, 6.2.

Eisenhardt, A., Knu, A., Schmidt, R., Himmerlich, M., Wagner, J., Schaefer, J. A., & Krischok, S. (2010). Phys. Status Solidi A 207, 1335. Fig. 5.24.

Eisenhardt, A., Eichapfel, G., Himmerlich, M., Knübel, A., Passow, T., Wang, C., Benkhelifa, F., Aidam, R., & Krischok, S. (2012). Phys. Status Solidi C 9, 685. Fig. 5.24.

Ejderha, K., Yıldırım, N., Türüt, A., & Abay, B. (2012). Eur. Phys. J. Appl. Phys. 57, 10102. Fig. 3.12.

Ejderha, K., Yıldırım, N., & Türüt, A. (2014). Eur. Phys. J. Appl. Phys 68, 20101. Fig. 5.19.

Elsbergen, V. van (1998). Ph. D. Thesis, Universität Duisburg Essen. Fig. 5.15.

Elsbergen, V. van, Kampen, T. U., & Mönch, W. (1996). J. Appl. Phys. 79, 316. Figs. 5.2, 5.15.

Emde, M. von der, Zahn, D. R. T., Schultz, C., Evans, D. A., & Horn, K. (1992). J. Appl. Phys. 72, 4486. Fig. 5.4.

Evans, D. A., Chen, T. P., Chasseé, T., Horn, K., Emde, M. von der, & Zahn, D. R. T. (1992). Surf. Sci. 269/270, 979. Fig. 5.4

Ewing, D. J., Porter, L. M., Wahab, Q., Ma, X., Sudharshan, T. S., Tumakha, S., Gao, M., & Brillson, L. J. (2007). J. Appl. Phys. 101, 114514. Fig. 5.15.

Eyckeler, M., Mönch, W., Kampen, T. U., Dimitrov, R., Ambacher, O., & Stutzmann, M. (1998). J. Vac. Sci. Technol. B 16, 2224. Fig. 5.19.

Fan, H. B., Sun, G. S., Yang, S. Y., Zhang, P. F., Zhang, R. Q., Wei, H. Y., Jiao, C. M., Liu, X. L., Chen, Y. H., Zhu, Q. S., & Wang, Z. G. (2008) Appl. Phys. Lett. 92, 192107. Fig. 5.16.

Fang, P., Rao, C., Liao, C., Chen, S., Wu, Z., Lu, X., Chen, Z., Wang, G., Liang, J., i Pei, Y. (2022). Semicond. Sci. Technol. 37, 115007. Fig. 5.25.

Fares, C., Ren, F., & Pearton, S. J. (2019a). ECS J. Solid State Sci. Technol. 8, Q3007. Fig. 5.25.

Fares, C., Tadjer, M. J., Woodward, J., Nepal, N., Mastro, M. A., Eddy Jr., C. R., Ren, F., & Pearton, S. J. (2019b). ECS J. Solid State Sci. Technol. 8, Q3154. Figs. 5.26, 5.27.

Farzana, E., Zhang, Z., Paul, P. K., Arehart, A. R., & Ringel, S. A. (2017). Appl. Phys. Lett. 110, 202102. Fig. 5.25.

Feng, Q., Feng, Z., Hu, Z., Xing, X., Yan, G., Zhang, J., Xu, Y., Lian, X., & Hao, Y. (2018). Appl. Phys. Lett. 112, 072103. Fig. 5.25.

Fleming, J. A. (1890). Proc. Roy. Soc. Lond. 47, 118. Sec. 1.1.

Fleming, J. A. (1905a). Proc. Roy. Soc. Lond. 74, 476. Sec. 1.1.

Fleming, J. A. (1905b). Brit. Patent 24850. Sec. 1.1.

Flores, F., Duran, J. C., & Muiiozt, A. (1987). Phys. Scr. T19, 102. Sects. 3.7, 3.8.2.

Foyt, A. G., Lindley, W. T., & Donnelly, J. P. (1970). Appl. Phys. Lett. 16, 335. Figs. 6.1, 6.2.

Freeouf, J. L., & Woodall, J. M. (1981). Appl. Phys. Lett. 39, 727. Sect. 2.4.

Freeouf, J. L., Jackson, T. N., Laux, S. E., & Woodall, J. M. (1982). Appl. Phys. Lett. 40, 634. Sect. 2.4, Fig. 2.10.

Fritsch, O. (1935). Ann. Physik 22, 375. Sect. 1.1.

Fritts, C. E. (1883). Amer. J. Sci. 26, 465. Sect. 1.1.

Fu, H., Baranowski I., Huang, X., Chen, H., Lu, Z., Montes, J., Zhang, X., & Zhao, Y. (2017). IEEE Ellectron Device Lett. 38, 1286. Fig. 5.21.

Fu, H., Chen, H., Huang, V., Baranowski, I., Montes, J., Yang, T. -H., & Zhao, Y. (2018). IEEE Trans. Electron Devices 65, 3507. Fig. 5.25.

Fujii, T., Shimomoto, K., Ohba, R., Toyoshima, Y., Horiba, K., Ohta, J., Fujioka, H., Oshima, M., Ueda, S., Yoshikawa, H., & Kobayashi, K. (2009). Appl. Phys. Express 2, 011002. Figs. 5.22, 5.23, 5.24.

Fulton, C. C., Lucovsky, G., & Nemanich, R. J. (2002). J. Vac. Sci. Technol. B 20, 1726. Fig. 5.12.

Fulton, C. C., Lucovsky, G., & Nemanich, R. J. (2006). J. Appl. Phys. 99, 063708. Figs. 5.12, 5.31, 5.32.

Gao, K. Y., Seyller, Th., Ley, L., Ciobanu, F., Pensl, G., Tadich, A., Riley, J. D., & Leckey, R. G. C. (2003). Appl. Phys. Lett. 83, 1830. Figs. 5.16, 5.29.

Gao, K. Y., Seyller, Th., Ley, L., Ciobanu, F., Pensl, G., Tadich, A., Riley, J. D., & Leckey R. G. C. (2003). Appl. Phys. Lett. 83, 1830. Fig. 5.29.

Garcia-Moliner, F., & Flores, F. (1979). *Introduction to the Theory of Solid Surfaces* (Cambridge Univ. Press, Cambridge). Sect. 3.6.

Geogakilas, A., Aperathitis, E., Foukaraki, V., Kayambaki, M., & Panayotatos, P. (1997). Mat. Sci. Engineer. B, 44, 383. Fig. 5.12.

Gerasimov, A. B. (1978). Sov. Phys. Semicond. 12, 709. Figs. 6.1, 6.2.

Ghosh, S., Baral, M., Kamparath, R., Singh, S. D., & Tapas Ganguli, T. (2019). Appl. Phys. Lett. 115, 251603. Figs. 5.26, 5.27.

Ghosh, A., Jana, S., Rauch, T., Tran, F., Marques, M. A. L., Botti, S., Constantin, L. A., Niranjan, M. K., & Samal, P. (2022). J. Chem. Phys. 157, 124108. Sect. 1.2, 4.1. Fig. 4.10.

Gibbon, J. T., Jones, L., Roberts, J. W., Althobaiti, M., Chalker, P. R., Mitrovic, I. Z., & Dhanak, V. R. (2018). AIP Advances 8, 065011. Figs. 5.12, 5.26, 5.27.

Giannazzo, F., Roccaforte, F., & Raineri, V. (2007). Microelectronic Engin. 84, 450. Fig. 5.15.

Giustino, F., Pasquarello, A. (2005). Surf. Sci. 586, 183. Tab. 4.2.

Godby, R. W., Schlüter, M., & Sham, L. J. (1988). Phys. Rev. B 37, 10159. Sect. 4.1.

Golan, A., Bregman, J., Shapira, Y., & Eizenberg, M. (1991). J. Appl. Phys. 69, 1494. Fig. 5.18.

Gong, M., Fung, S., Beling, C. D., & You, Z. (1999). J. Appl. Phys. 85, 7120. Fig. 6.3.

Gong, Y. -P., Li, Ai-D., Liu, X. -J., Zhang, W. -Qi, Li, H., & Wu, Di (2011). Surf. Interface Anal. 43, 734. Fig. 5.29.

Goniakowski, J., & Noguera, C. (2004). Interface Sci. 12, 93. Chapt. 7.

Goodman, A. M. (1968), Appl. Phys. Lett. 13, 275. Fig. 5.33.

Goodman, A. M., & O'Neill, J. J. (1966). J. Appl. Phys. 37, 3580. Fig. 5.30.

Goodman, S., Auret, F. D., Deenapanray, P. N. K., & Myburg, G. (1998). Jpn. J. Appl. Phys. 37, L10. Fig. 5.5.

Goodwin, E. T. (1939). Proc. Cambr. Philos. Soc. 35, 205. Sect. 3.6.

Gordy, W., & Thomas, W. J. O. (1956). J. Chem. Phys. 24, 439. Sect. 3.3.

Grey, F., Feidenhans'l, R., Nielsen, M., & Johnson, R. L. (1989). J. Phys. Coll. C7, 181. Sect. 2.5.

Gritsenko, V. A., Shaposhnikova, A. V., Kwokb, W. M., Wongc, H., & Jidomirovd, G. M. (2003) Thin Solid Films. 437, 135. Fig. 5.33.

Grodzicki, M., Mazur, P., Zuber, S., Brona, J., & Ciszewski, A. (2014). Appl. Surf. Sci. 304, 20. Figs. 5.20, 5.26, 5.27.

Grondahl, L. O. (1925), U. S. Patent No. 1,640,335. Sect. 1.1.

Grondahl, L. O. (1926), Phys. Rev. 26, 813. Sect. 1.1.

Grondahl, L. O. (1933), Rev. Mod. Phys. 5, 141. Sect. 1.1.

Grunthaner, F. J., Grunthaner, P. J. (1986). Mater. Sci. Rep. 1, 65. Figs. 5.12, 5.31, 5.32.

Grunwald, F. (1987) Diploma Thesis, Universität Duisburg (unpublished), Fig. 5.17.

Gullu, H. H., Sirin, D. S., & Yıldız, D. F. (2021). J. Electron. Mater. 50, 7044. Fig. 5.15.

Guo, Y., Liu, X. -L., Song, H. -P., Yang, An-Li, Zheng, G. -L., Wei, H. -Y., Yang, S. -Y., Zhu, Q. -S., & Wang, Z. -G. (2010). Chin. Phys. Lett. 27, 067302. Fig. 5.13.

Guo, Q., Takahashi, K., Saito, K., Akiyama, H., Tanaka, T., & Nishio, M. (2013). Appl. Phys. Lett. 102, 092107. Fig. 5.18.

Guo, Y., Li, H., Clark, S. J., & Robertson J. (2019). J. Phys. Chem. C 123, 5562. Sects. 1.2, 4.1. Fig. 4.9.

Gurnett, M., Gustafsson, J. B., Holleboom, L. J., Magnusson, K. O., Widstrand, S. M., & Johansson L. S. O. (2005). Phys. Rev. B 71, 195408. Fig. 3.10.

Hacke, P., Detchprohm, T., Hiramatsu, K., & Sawaki, N. (1993). Appl. Phys. Lett. 63, 2676. Fig. 5.19.

Hagemann, U., Huba, K., & Nienhaus, H. (2018). Appl. Phys. 124, 225302. Figs. 2.18, 5.1.

Hanney, N. B., & Smith, C. P. (1946). J. Am. Chem. Soc. 68, 171. Sect. 3.5.1.

Hansen, P. J., Vaithyanathan, V., Wu, Y., Mates, T., Heikman, S., Mishra, U. K., York, R. A., Schlom. D. G., & Speck, J. S. (2005). J. Vac. Sci. Technol. B 23, 499. Fig. 5.20.

Harada, T., Ito, S., & Tsukazaki, A. (2019). Sci. Adv. 5, eaax5733. Fig. 5.25.

Hasegawa, H., & Ohno, H. (1986). J. Vac. Sci. Technol. B 4, 1130. Sect. 1.2.

Hashizume, T., Ootomo, S., & Hasegawa, H. (2003). Appl. Phys. Lett. 83, 2952. Fig. 5.29.

Hattori, M., Oshima, T., Wakabayashi, R., Yoshimatsu, K., Sasaki, K., Masui, T., Kuramata, A., Yamakoshi, S., Horiba, K., Kumigashira, H., & Ohtomo, A. (2016). Jpn. J. Appl. Phys. 55, 1202B6. Figs. 5.26, 5.27, 5.29.

He, Q., Mu, W., Dong, H., Long, S., Jia, Z., Lv, H., Liu, Qi, Tang, M., Tao, X., & Liu, M. (2017). Appl. Phys. Lett. 110, 093503. Fig. 5.25.

Heine, V. (1965). Phys. Rev.138, A1689. Sects. 1.2, 1.3, 3.6, 3.7, 3.8.2, 4.1.

Helal, H., Benamara, Z., Arbia, M. B., Khettou, A., Rabehi, A., Kacha, A. H., & Amrani, M. (2020). Int. J. Numer. Model El. 2714. Fig. 5.19.

Hemmingson, C., Son, N. T., Kordina, O., Bergman, J. P., Janzén, E., Lindström, J. L., Savage, S., & Nordell, N. (1997). J. Appl. Phys. 81, 6155. Fig. 6.3.

Higashiwaki, M., Sasaki, K., Goto, K., Nomura, K., Thieu. Q. T., Togashi, R., Murakami, H., Kumagai, Y., Monemar, Bo, Koukitu, A., Kuramata, A., & Yamakoshi, S. (2015). 73rd Annual Device Research Conference (DRC), p.29. Fig. 5.25.

Higuchi, M., Sugawa, S., Ikenaga, E., Ushio, J., Nohira, H., Maruizumi, T., Teramoto, A., Ohmi, T., & Hattori, T. (2007). Appl. Phys. Lett. 90, 123114. Fig. 5.33.

Himpsel, F. J., McFeely, F. R., Taleb-Ibrahimi, A., Yarmoff, J. A., & Hollinger, G. (1988). Phys. Rev. B 38, 6084. Figs. 5.12, 5.31, 5.32.

Hinuma, Y., Grüneis, A., Kresse G., & Oba, F. (2014). Phys. Rev. B 90, 155405. Sects. 4.1, 4.2. Fig. 4.6.

Hirose, K., Sakano, K., Nohira, H., & Hattori, T. (2001). Phys. Rev. B 64, 155325. Figs. 5.12, 5.31, 5.32.

Höffling, B., Schleife, A., Rödl, C., & Bechstedt F. (2012). Phys. Rev. B 85, 035305. Sect. 1.2, 4.1, 4.2, 5.2.11. Fig. 4.5.

Hong, H., Aburano, R. D., Lin, D. -S., Chen, H., & Chiang, T. -C. (1992). Phys. Rev. Lett. 68, 507. Sects. 1.2, 2.5.

Hong, S. -Ku, Hanada, T., Makino, H., Chen, Y., Ko, H. -Ju, & Yao, T. (2001). Appl. Phys. Lett. 78, 3349. Fig. 5.20.

Horváth, Zs. J., Ayyildiz, E., Rakovics, V., Cetin, H., & Põdör, B. (2005). physica status solidi 2, 1423. Fig. 2.22.

Howes, P. B., Edwards, K. A., Hughes, D. J., Macdonald, J. E., Hibma, T., Bootsma, T., & James, M. A. (1995). Phys. Rev. B 51, 17740. Sects. 1.2.

Hou, C., Gazoni, R. M., Reeves, R. J., & Allen, M. W. (2019a). Appl. Phys. Lett. 114, 033502. Fig. 5.25.

Hou, C., Gazoni, R. M., Reeves, R. J., & Allen, M. W. (2019b). IEEE Electron Device Lett. 40, 337. Fig. 5.25.

Hou, C., Gazoni, R. M., Reeves, R. J., & Allen, M. W. (2019c). IEEE Electron Device Lett. 40, 1587. Fig. 5.25.

Hu, Z., Feng, Q., Feng, Z., Cai, Y., Shen, Y., Yan, G., Lu, X., Zhang, C., Zhou, H., Zhang, J., & Hao, Y. (2019). Nanoscale Res. Lett. 14, 2. Fig. 5.25.

Hu, Z., Li, J., Zhao, C., Feng, Z., Tian, X., Zhang, Y., Zhang, Y., Ning, J., Zhou, H., Zhang, C., Lv, Y., Kang, X., Feng, H., Feng, Q., & Zhang, J. (2020). IEEE Transact. Electron Devices 67, 5628. Figs. 5.26, 5.27.

Huan, Ya-W., Wang, X. -Lu, Liu, W. -J., Dong, H., Long, S. -B., Sun, S. -M., Yang, J. -G., Wu, Su-D., Yu, W. -J., Horng, R. -H., Xia, C. -T., Yu, H. -Yu, Lu, H. -L., Sun, Q. -Q., Ding, S. -J., & Zhang, D. -W. (2018). Jpn. J. Appl. Phys. 57, 100312. Figs. 5.26, 5.27.

Huang, L., & Wang, D. (2015). Jpn. J. Appl. Phys. 54, 114101. Fig. 5.15.

Huang, M. L., Chang, Y. C., Chang, C. H., Lin, T. D., Kwo, J., Wu, T. B., & Hong, M. (2006). Appl. Phys. Lett. 89, 012903. Fig. 5.29.

Huang, W. -C., Horng, C. -T., Chen, Yu-M., Chen, C. -C. (2010). Phys. Status Solidi C 7, 2326. Fig. 5.6.

Huba, K., Krix, D., Meier, C., & Nienhaus, H. (2009). J. Vac. Sci. Technol. A 27, 889. Figs. 2.18, 5.1.

Hudait, M. K., & Zhu, Y. (2013). J. Appl. Phys. 113, 114303. Fig. 5.13.

Hübers, H. -W., & Röser, H. P. (1998). J. Appl. Phys. 84, 5326. Fig. 5.17.

Ibach, H., & Lüth, H. (1991). *Festkörperphysik – Einführung in die Grundlagen* (Springer, Berlin), Sect. 3 6.

Ihm, J., Louie, S. G., & Cohen, M. L. (1978). Phys. Rev. Lett. 40, 1208. Tab. 4.2.

Im, H. -J., Kaczer, B., Pelz, J. P., & Choyke, W. J. (1998). Appl. Phys. Lett. 72, 839 and private communication. Figs. 2.23, 515.

Im, H. -J., Ding, Y., Pelz, J. P., & Choyke, W. J. (2001). Phys. Rev. B 64, 075310. Fig. 5.15.

Irmscher K., Galazka, Z., Pietsch, M., Uecker,R., & Irmscher, R. F. (2011). J. Appl. Phys. 110, 063720. Fig. 5.25.

Iucolano, F., Roccaforte, F., Giannazzo, F., & Raineri, V. (2007). Appl. Phys. Lett. 90, 092119. Fig. 5.19.

Izzo, G., Litrico, G., Severino, A., Foti, G., La Via, F., & Calcagno, L. (2009). Mater. Sci. Forum 615–617, 397. Fig. 6.3.

Jacob, W., Bertel, E., & Dose, V. (1987). Europhys. Lett. 4, 1303. Sect. 1.2.

Jäger, D., & Kassing R. (1977). J. Appl. Phys. 48, 4413. Figs. 3.11, 5.11.

Jain, N., Zhu, Y., Maurya, D., Varghese, R., Priya, S., & Hudait, M. K. (2014). J. Appl. Phys. 115, 024303. Fig. 5.13.

Jeon, K. -R., Lee, S. -J., Park, C. -Y., Lee, H. -S., & Shin, S. -C. (2010). Appl. Phys. Lett. 97, 111910. Fig. 5.13.

Ji, X., Yue, J., Qi, X., Meng, D., Chen, Z., & Li, P. (2021). J. Appl. Phys. 130, 075301. Figs. 5.26, 5.27.

Jia, C. H., Chen, Y. H., Zhou, X. L., Yang, A. L., Zheng, G. L., Liu, X. L., Yang, S. Y., & Wang, Z. G. (2009). J. Phys. D 42, 095305. Fig. 5.34.

Jia, Ye, Zeng, K., Wallace, J. S., Gardella, J. A., & Singisetti, U. (2015a). Appl. Phys. Lett. 106, 102107. Figs. 5.26, 5.27.

Jia, Ye, Wallace, J. S., Qin, Y., Gardella JR., J. A., Dabiran, A. M., & Singisetti, U. M. (2015b). J. Electron Mater. 45, 2013. Fig. 5.24.

Jia, Ye, Wallace, J. S., Echeverria, E., Gardella Jr., J. A., & Singisetti, U. (2017). Phys. Status Solidi B 254, 1600681. Fig. 5.20.

Jian, G.,He, Q., Mu, W., Fu, B., Dong, H., Qin, Y., Zhang, Y., Xue, H., Long, S., Jia, Z., Lv, H., Liu, Q., Tao, X., Liu, M. (2018). AIP Adv. 8, 015316 Fig. 5 25

Jian, Z., Mohanty, S., & Ahmadi, E. (2020). Appl. Phys. Lett. 116, 152104. Fig. 5.25.

Jin, H., Oh, S. K., Kang, H. J., Lee, Y. S., & Cho, M. H. (2006). Surf. Interface Anal. 38, 502. Fig. 5.12.

Johnson, W. E., & Lark-Horovitz, K. (1949). Phys. Rev. 76, 442. Chapt. 6.

Joishi, C., Rafique, S., Xia, Z., Han, Lu, Krishnamoorthy, S., Zhang, Y., Lodha, S., Zhao, H., & Rajan, S. (2018). Appl. Phys. Express 11, 031101. Fig. 5.25.

Jyothi, I., Reddy, V. R., Reddy, M. S . P., Choi, C. -J, & Bae, J. -S. (2010). Phys. Status Solidi A 207, 753. Fig. 5.19.

Käckell, P., Wenzien, B., & Bechstedt, F. (1994). Phys. Rev. B 50, 10761. Sect. 5.1.1.

Kalinina, E. V., Kuznetsov, N.I., Babanin, A.I., Dmitriev, V. A., & Shchukarev , A. V. (1997). Diamond Relat. Mater. 6, 1528. Fig. 5.19.

Kamimura, T., Sasaki, K., Wong, M. H., Krishnamurthy, D., Kuramata, A., Masui, T., Yamakoshi, S., & Higashiwaki, M. (2014). Appl. Phys. Lett. 104, 192104. Figs. 5.26, 5.27, 5.29.

Kampen, T. U., & Mönch, W. (1995). Surf. Sci. 331, 490. Fig. 2.18.

Kampen, T. U., & Mönch, W. (1997). Appl. Surf. Sci. 117, 388. Fig. 5.19.

Karatas, S., Altindal, A., Türüt, A., & Özmen, A. (2003). Appl. Surf Sci. 217, 250. Fig. 2.18.

Karoui, M. B., Gharbi, R., Alzaied, N., FathallahM., Tresso, E., Scaltrito, L., & Ferrero, S. (2008). Mater. Sci. Engin. C 28, 799. Fig. 5.15.

Karrer, U., Dobner, A., Ambacher, O., & Stutzmann, M. (1999). Phys. Status Solidi A 176, 163. Fig. 5.19.

Kato, M., Ichimura, M., Arai, E., Masuda, Y., Chen, Yi, Nishino, S., & Tokuda, Y. (2001). *Jpn. J. Appl. Phys.* 40, 4943. Fig. 6.3.

Kavaliunas, V, Hatanaka, Y., Neo, Y., Laukaitis, G., & Mimura, H. (2021). ECS J. Solid State Sci. Technol. 10, 015005. Figs. 3.11, 5.11, 5.12.

Kawahara, K., Alfieri, G., & Kimoto, T. (2009). J. Appl. Phys. 106, 013719. Fig. 6.3.

Kawarada, H., Aoki, M., Sasaki, H., Tsugawa, K., & Ohdomari, I. (1994). in *Control of Semiconductor Interfaces,* edited by I. Ohdomari, M. Oshima, and A. Hiraki (Elsevier, Amsterdam), p. 161. Sects. 2.5.2, 5.3. Chapt. 7. Fig. 2.18.

Keister, J. W., Rowe, J. E., Kolodziej, J. J., Niimi, H., Madey, T. E., & Lucovsky, G. (1999). J. Vac. Sci. Technol. B 17, 1831. Figs. 5.12, 5.31, 5.32, 5.33.

Khan, M. R. H., Detchprohmt, T., Hackez, P., Hiramatsut, K., & Sawakiz, N. (1995). J. Phys. D: Appl. Phys. 28, 1169. Figs. 5.19, 5.20.

Khemka, V., Chow, T. P., Gutmann, R. J. (1998). J. Electron. Mater. 27, 1128. Fig. 5.15.

Kim, H. (2010). Mater. Sci. Semicond. Processing 13, 51. Fig. 5.20.

Kim, H. (2024). Brazil. J. Phys. 54, 34. Fig. 5.19.

Kim, J., Pearton, S. J., Fares, C., Yang, J., Ren, F., Kima, S., & Polyakov, A. Y. (2019). J. Mater. Chem. C 7, 10. Chapt. 6. Fig. 6.5.

Kim, D. G., Kim, H. -R., Kwon, D. S., Lim, J., Seo, H., Kim, T. K., Paik, H., Lee, W., & Hwang, C. S. (2021). J. Phys. D - Appl. Phys. 54, 185110. Fig. 5.13.

King, S. W., Ronning, C., Davis, R. F., Benjamin, M. C., & Nemanich, R. J. (1998). J. Appl. Phys. 84, 2086. Fig. 5.20, 5.22, 5.23.

King, S. W., Davis, R. F., Ronning, C., & Nemanich, R. J. (1999). J. Electron. Mater. 28, L34. Figs. 5.16, 5.20.

King, P. D. C., Veal, T. D., Jefferson, P. H., McConville, C. F., Wang, T., Parbrook, P. J., Lu, H., & Schaff, W. J. (2007). Appl. Phys. Lett. 90, 132105. Fig. 5.20, 5.22, 5.23, 5.24.

King, P. D. C., Veal, T. D., Kendrick, C. E., Bailey, L., Durbin, S. M., & McConville, C. F. (2008a). Phys. Rev. B 78, 033308. Fig. 5.20.

King, P. D. C., Veal, T. D., Jefferson, P. H., Hatfield, S. A., Piper, L. F. J., McConville, C. F., Fuchs, F., Furthmüller, J., Bechstedt, F., Lu, H., & Schaff, W. J. (2008b). Phys. Rev. B 77, 045316. Fig. 5.24.

King, P. D. C., Veal, T. D., Jefferson, P. H., Zúñiga-Pérez, J., Muñoz-Sanjosé, V., & McConville, C. F. (2009). Phys. Rev. B 79, 035203. Figs. 6.1, 6.2.

King, S. W., Nemanich, R. J., & Davis, R. F. (2015a). J. Appl. Phys. 118, 045304. Figs. 3.11, 5.11, 5.15, 5.22, 5.23.

King, S. W., Nemanich, R. J., & Davis, R. F. (2015b). Phys. Status Solidi B 252, 391. Fig. 5.15.

Klein, A., Henrion, O., Pettenkofer, C., Jaegermann, W., Ashkenasy, N., Mishori, B., & Shapira, Y. (1997). Proc. of the 14th European Photovoltaic Solar Energy Conference, edited by H. S. Stephens, p. 1705. Fig. 5.18.

Krishna, S., & Gupta, G. (2014). RSC Adv. 4, 27308. Fig. 5.24.

Koenigsberger, J., & Weiss, J. (1911). Ann. Physik (IV) 35, 1. Sect. 1.1.

Kojima, K., Yoshikawa, M., Ohshima, T., Itho, H., & Okada, S. (2000). Mater. Sci. Forum 338–342, 1239. Fig. 5.2.

Konishi, K., Kamimura, T., Wong, M. H., Sasaki, K., Kuramata, A., Yamakoshi, S., & Higashiwaki, M. (2016). Phys. Status Solidi B, 253, 623. Figs. 5.26, 5.27.

Konozenkao, I. D., K, Semenyuk, K., & Khivrich, V.I. (1969). phys. status solidi (b) 35, 1048. Figs. 6.1, 6.2.

Korkut, H. (2013). Nano-Micro Lett. 5, 34. Fig. 3.12.

Korucu, D., & Duman, S. (2013) Thin Solid Films. 531, 436. Fig. 3.12.

Kowalczyk, S. P., Kraut, E. A., Waldrop, J. R., & Grant, R. W. (1982a). J. Vac. Sci. Technol. 21, 482. Fig. 5.18.

Kowalczyk, S. P., Schaffer, W. J., Kraut, E. A., & Grant, R. W. (1982b). J. Vac. Sci. Technol. 20, 705. Fig. 5.18.

Kraut, E. A., Grant, R. W., Waldrop, J. R., & Kowalczyk, S. P. (1980). Phys. Rev. Lett. 44, 1620. Sect. 3.8.3. Fig. 5.13, 5.18.

Kribes, Y., Harrison, I., Tuck, B., Cheng, T. S., & Foxon, C. T. (1997). Semicond. Sci. Technol. 12, 913. Fig. 5.19.

Krix, D., Huba, K., & Nienhaus, H. (2009). J. Vac. Sci. Technol. A 27, 918. Fig. 2.18.

Kroemer, H. (1957). Proc. IRE 45, 1535; RCA Rev. 28, 332. Sect. 1.3.

Kroemer, H. (1963). Proc. IEEE 51, 1782. Sect. 1.3.

Kroemer, H. (1975). Crit. Revs. Solid State Mater. Sci. 5, 555. Sect. 3.8.2.

Kroemer, H. (1985). in *Molecular Beam Epitaxy and Heterostructures*, edited by L. L. Chang and K. Ploog (M. Nijhoff, Dordrecht), p. 331. Sect. 3.8.2.

Kroemer, H. (2001). Rev. Mod. Phys. 73, 783. Sect. 1.3.

Kuang, Y., Chen, X., Ma, T., Du, Q., Zhang, Y., Hao, J., Ren, F. -F., Liu, B., Zhu, S., Gu, S., Zhang, R., Zheng, Y., & Ye, J. (2021). ACS Appl. Electron. Mater. 3, 795. Figs. 5.26, 5.27.

Kumar, A. A., Rao, L. D., Reddy, V. R., & Choiet, Ch. -J. (2013). Current Appl. Phys. 13, 975. Fig. 3.12.

Kumar, M., Roul, B., Bhat, T. N., Rajpalke, M. K., Kalghatgi, A. T., & Krupanidhi, S. B. (2012). Thin Solid Films 520, 4911. Figs. 5.33, 7.4.

Kumar, A. A., Sharma, K. K., & Chand, S. (2019). Superlatt. Microstruc. 128, 373. Sect. 2.4.

Kuo, C. -T., Chang, K. -K., Shiu, H. -W., Liu, C. -R., Chang, Lo-Y., Chen, C. -H., & Gwo, S. (2011a). Appl. Phys. Lett. 99, 122101. Fig. 5.12.

Kuo, C. -T., Chang, K. -K., Shiu, H. -W., Lin, S. -C., Chen, C. -H., & Gwo, S. (2011b). Appl. Phys. Lett. 99, 022113. Figs. 5.22, 5.23, 5.24.

Kurtin, S., & Mead, C. A. (1969). J. Phys. Chem. Solids 30, 2007. Fig. 5.6.

Kurtin, S., McGill, T. C., & Mead, C. A. (1969). Phys. Rev. Lett. 22, 1433. Sect. 1.2.

Kusaka, M., Hiraoka, N., Hirai, M., & Okazaki, S. (1980). Jpn. J. Appl. Phys. 19, 1187. Fig. 5.4.

Kuzel, R., & Weidmann, F. L. (1970). Can. J. Phys. 48, 2643. Sect. 1.1.

Lang, D. V. (1974). J. Appl. Phys. 45, 3023. Chapt. 6.

Lark-Horovitz, K., Bleuler, E., Davis, R., & Tendam, D. (1948). Phys. Rev. 73, 1256. Chapt. 6.

Lebedev, A. A., Veĭnger, A.I., Davydov, D. V., Kozlovskiĭ, V. V., Savkina, N. S., & Strel'chuk, A. M. (2000a). Semiconductors 34, 861. Fig. 6.3.

Lebedev, A. A., Davidov, D. V., Strel'chuk, A. M., Kuznetzov, A. N., Bogdanova, E. V., Kozlovskiĭ, V. V., & Savkina, N. S. (2000b). Mater. Sci. Forum 338–342, 973. Fig. 6.3.

Lebedev, A. A., Veĭnger, A.I., Davydov, D. V., Kozlovskiĭ, V. V., Savkina, N. S., & Strel'chuk, A. M. (2000c). Semiconductors 34, 1016.

Lee, S. -K., Zetterling, C. -M., & Östling, M. J. (2001). Electron. Mater. 30, 242. Fig. 5.2

Lee, S. -Y., Kim, T. -H., Cho, N. -K., Seong, H. -K., Choi, H. -J., Ahn, B. -G., & Lee, S. -K. (2008). J. Nanosci. Nanotechnol. 8, 5042. Fig. 5.20.

Lee, K., Nomura, K., Yanagi, H., Kamiya, T., Ikenaga, E., Sugiyama,T., Kobayashi, K., & Hosono, H. (2012). J. Appl. Phys. 112, 033713. Fig. 7.4.

Lei, M., Yum, J. H., Banerjee, S. K., Bersuker, G., & Downer, M. C. (2012). Phys. Status Solidi B 249, 1160. Fig. 5.12.

Leroy, W. P., Opsomer, K., Forment, S., & van Meirhaeghe, R. S. (2009). Solid-State Electron. 49, 878. Fig. 5.17.

Lewandków, R., Grodzicki, M., Mazur, P., & Ciszewski, A. (2021). Surf. Interface Anal. 53, 118. Fig. 5.20.

Li, S. X., Yu, K. M., Wu, J., Jones, R. E., Walukiewicz, W., Ager III, J. W., Shan, W., Haller, E. E., Lu, H., & Schaff, W. J. (2005). Phys. Rev. B 71, 161201. Figs. 6.1, 6.2.

Li, Y., Long, W., & Tung, R. T. (2011a). Solid State Commun. 151, 1641. Sect. 2.5.2.

Li, Z., Zhang, B., Wang, J., Liu, J., Liu, X., Yang, S., Zhu, Q., & Wang, Z. (2011b). Nanoscale Res. Lett. 6, 193. Figs. 5.24, 5.34.

Li, H., Liu, X., Sang, L., Wang, J., Jin, D., Zhang, H., Yang, S., Liu, S., Mao, W., Hao, Y., Zhu, Q., & Wang, Z. (2014). Phys. Status Solidi B 251, 788. Figs. 5.22, 5.23.

Li, K. -H., Alfaraj, N., Kang, C. H., Braic, L., Hedhili, M. N., Guo, Z., Ng, T. K., & Ooi, B. S. (2019). ACS Appl. Mater. Interfaces 11, 35095. Figs. 5.26, 5.27.

Li, W., Zhang, X., Zhao, J., Yan, J., Liu, Z., Wang, J., Li, J., & Wei T. (2020a). J. Appl. Phys. 127, 015302. Figs. 5.26, 5.27.

Li, K. -H., Kang, C. H., Min, J. -H., Alfaraj, N., Liang, J. -W., Braic, L., Guo, Z., Hedhili, M. N., Ng, T. K., & Ooi, B. S. (2020b). ACS Appl. Mater. Interfaces 12, 53932. Figs. 5.26, 5.27.

Liang, Y., Curless, J., & McCready, D. (2005). Appl. Phys. Lett. 86, 082905. Fig. 5.18, 5.34.

Lieten, R. R., Degroote, S., Kuijk, M., & Borghs, G. (2008). Appl. Phys. Lett. 92, 022106. Fig. 5.13.

Lim, Z. H., Ahmadi-Majlan, K., Grimley, E. D., Du, Y., Bowden, M., Moghadam, R., LeBeau, J. M., Chambers, S. A., & Ngai, J. H. (2017). J. Appl. Phys. 122, 084102. Fig. 5.13.

Lin, Y. -J. (2009). J. Appl. Phys.106, 013702. Figs. 5.19, 5.20.

Lin, D. -S., Miller, T., & Chiang, T. -C. (1991). Phys. Rev. B 44, 10719. Sect. 3.5.2.

Lin, J. -Y., Roy, A. M, Nainani, A., Sun, Y., & Saraswat, K. C. (2011). Appl. Phys. Lett 98, 092113. Fig. 5.29.

Lingaparthi, R., Thieu, Q. T., Koshi, K., Wakimoto, D., Sasaki, K., & Kuramata A. (2020). Appl. Phys. Lett. 116, 152104. Fig. 5.25.

Liu, Q. Z., Yu, L. S., Deng, F., Lau, S. S., & Redwing, J. M. (1998). J. Appl. Phys. 84, 881. Fig. 5.19.

Liu, H. F., Hu, G. X., Gong, H., Zang, K. Y., & Chua, S. J. (2008). J. Vac. Sci. Technol. A 26, 1462. Fig. 5.20.

Liu, J. W., Kobayashi, A., Ueno, K., Toyoda,S., Kikuchi, A., Ohta, J., Fujioka, H., Kumigashira, H., & Oshima, M. (2010). Appl. Phys. Lett. 97, 252111. Figs. 5.22, 5.23.

Liu, J. W., Kobayashi, A.,Toyoda, S., Kamada, H., Kikuchi, A., Ohta, J., Fujioka, H., Kumigashira, H., & Oshima, M. (2011a). Phys. Status Solidi B 248, 956. Fig. 5.20.

Liu, Z. Q., Chim, W. K., Chiam, S. Y., Pan, J. S., & Ng, C. M. (2011b). J. Appl. Phys. 109, 093701. Fig. 7.4.

Liu, Z. Q., Chim, W. K., Chiam, S. Y., Pan, J. S., & Ng, C. M. (2012). J. Mater. Chem. 22, 17887. Fig. 7.4.

Liu, J. W., Liao, M. Y., Cheng, S. H., Imura, M., & Koide Y. (2013a). J. Appl. Phys. 113, 123706. Fig. 3.10.

Liu, J., Kobayashi, A., Ueno, K., Ohta, J., Fujioka, H., & Oshima, M. (2013b). Jpn. J. Appl. Phys. 52, 011101. Figs. 5.31, 5.32.

Liu, Z., Liu, Y., Wang, X., Li, W., Zhi, Y., Wang, X., Li, P., & Tang, W. (2019a). J. Appl. Phys. 126, 045707. Figs. 5.26, 5.27.

Liu, Z., Yu, J., Li, P., Wang, X., Zhi, Y., Chu, X., Wang, X., Li, H., Wu, Z., & Tang, W. (2019b). J. Phys. D - Appl. Phys. 52, 295104. Fig. 5.29.

Liu, H., Liu, W. -J., Xiao, Yi-F., Liu, C. -C., Wu, X. -H., & Ding, S. -J. (2020). Chin. Phys. Lett. 37, 077302. Figs. 5.26, 5.27.

Liu, Y., Wang, L., Zhang, Y., Dong, X. Sun, X., Hao, Z., & Luo, Yi (2021). IEEE Electron Device Lett. 42, 409. Figs. 5.26, 5.27.

Long, W., Li, Y., & Tung, R. T. (2013). Surf. Sci. 610, 48. Sect. 2.5.2, Fig. 3.11.

Long, W., Li, Y., & Tung, R. T. (2014) Thin Solid Films. 557, 254. Sect. 2.5.2.

Louie, S. G, & Cohen, M. L. (1976). Phys. Rev. B 13, 2461. Sects. 1.2, 3.6. Tab. 4.2.

Louie, S. G, Chelikowsky, J. R., & Cohen, M. L. (1977). Phys. Rev. B 15, 2154. Sect. 3.6, 4.2.

Louis, E., Yndurain, F., & Flores, F. (1976). Phys. Rev. B 13, 4408. Sect. 4.1. Tab. 4.2.

Lu, Y., Le Breton, J. C., Turban, P., Lépine, B., Schieffer, P., & Jézéquel. B. (2006a). Appl. Phys. Lett. 88, 042108. Fig. 5.18.

Lu, Y., Assi, C. K., Le Breton, J. C., Turban, P., Lépine, B., Schieffer, P., & Jézéquel, G. (2006b). J. Phys. IV France 132, 63. Tabl. 5.2, Chapt. 7.

Lu, X., Zhou, X., Jiang, H., Ng, K. W., Chen, Z., Pei, Y., Lau, K. M., & Wang, G. (2020). IEEE Electron Device Lett. 41, 449. Figs. 5.26, 5.27.

Lu, C., Gao, L., Meng, F., Zhang, Q., Yang, L., Liu, Z., Zhu, M., Chen, X., Lyu, X., Wang, Y., Liu, J., Ji, A., Li, P., Gu, L., Cao, Z., & Lu, N. (2023). J. Appl. Phys. 133, 045306. Figs. 5.26, 5.27, 5.34.

Lucovsky, G., Rayner Jr., G. B., Zhang, Y., Fulton, C. C., Nemanich, R. J., Appel, G., Ade, H., & Whitten, J. L. (2003). Appl. Surf. Sci. 212–213, 563. Figs. 5.12, 5.31, 5.32.

Lugakov, P. F., Lukashevich, T. A., V. V., & Shusha, V. V. (1982). phys. stat. sol. (a) 74, 445.

Lyle, L. A. M., Jiang, K., Favela, E. V., Das, K., Popp, A., Galazka, Z., Wagner, G., & Porter, L. M. (2021). J. Vac. Sci. Technol. A 39, 033202. Fig. 5.25.

Ma, P., Guo, W., Sun, J., Gao, J., Zhang, G., Xin, Q., Li, Y., & Song, A. (2019). Semicond. Sci. Technol. 34, 105004. Fig. 5.29.

Maeda, K., & Kitahara, E. (1998). Appl. Surf. Sci. 130, 925. Figs. 3.11, 5.11.

Maffeis, T. G. G., Simmonds, M. C., Clark, S. A., Peiro, P., Haines, P., & Parbrook, P. J. (2000). J. Phys. D: Appl. Phys. 33, L115. Fig. 5.20.

Magnusson, K. O., Wiklund, S., Dudde, R., & Reihl, B. (1991). Phys. Rev. B 44, 5657. Sect. 3.5.2.

Mahapatra, R., Chakraborty, A. K., Horsfall, A. B., Wright N. G., Beamson, G., & Coleman, K. S. (2008). Appl. Phys. Lett. 92, 042904. Fig. 5.16.

Mahato, S., & Puigdollers, J. (2028). Physica B: Phys. Condens. Matter 530, 327. Sects. 2.4, 2.6.1, Figs. 2.13, 2.14, 2.19.

Mahmood, Z. H., Shah, A. P., Kadir, A., Gokhale, M. R.., Ghosh, S., Bhattacharya, A., & Arora, B. M. (2007). Appl. Phys. Lett. 91, 152108. Figs. 5.20, 5.24.

Majdi, S., Gabrysch, M., Balmer, R., Twichen, D., & Isberg, J. (2010). J. Electronic Mater. 39, 1203. Figs. 2.18, 5.1.

Makimoto, T., Kumakura, K., Nishida, T., & Kobayashi, N. (2002). J. Electron. Mater. 31, 313. Fig. 5.20.

Makimoto, T., Kashiwa, M., Kido, T., Matsumoto, N., Kumakura, K., & Kobayashi, N. (2003). Phys. Stat. Sol. (c) 0, 2393. Fig. 5.19.

Mamor, M. (2009). J. Phys. - Condens. Matter 21, 335802. Figs. 5.19, 5.20.

Manago, T., Miyanishi, S., Akinaga, H., Van Roy, W., Roelfsema, R. F. B., Sato, T., Tamura, E., & Yuasa, S. (2000). J. Appl. Phys. 88, 2043. Fig. 5.17.

Maréchal, A., Kato, Y., Liao, M., & Koizumi, S. (2017). Appl. Phys. Lett. 111, 141605. Fig. 5.10.

Margaritondo, G., Stoffel, N. G., Katnani, A. D., Edelman, H. S., & Bertoni, C. M. (1981). J. Vac. Sci. Technol. 18, 784. Fig. 5.13.

Margaritondo, G., Katnani, A. D., Stoffel, N. G., Daniels, R. R., & Zho, T. -X. (1982). Solid State Commun. 43, 163. Fig. 5.12.

Martin, G., Strite, S., Botchkarev, A., Argawal, A., Rockett, A., Morkoç, H., Lambrecht, W. R. L., & Segal, B. A. (1994). Appl. Phys. Lett. 65, 610. Figs. 5.20, 5.22, 5.23.

Martin, G., Botchkarev, A., Rockett, A., & Morkoç, H. (1996). Appl. Phys. Lett. 68, 2541. Figs. 5.20, 5.22, 5.23. 5.24.

Mashovets, T. V., Khansevarov, & R. Yu (1966). Sov. Phys. -Solid State 8, 1350. Fig. 6.1.

Matsuo, N., Doko, N., Yasukawa, Y., Saito, H., & Yuasa, S. (2018). Jpn. J. Appl. Phys. 57, 070304. Figs. 5.26, 5.27.

Mattern, B., Bassler, M., Pensl, G., & Ley, L. (1998). Mater. Sci. Forum 264–268, 375. Figs. 5.16, 5.31, 5.32.

Maue, A. W. (1935). Z. Physik 94, 717. Sects. 1.2, 3.6, 4.1, 4.2.

Mazari, H., Benamara, Z., Bonnaud, O., & Olier, R. (2002). Mater. Sci. Engineering B 90, 171. Fig. 5.5.

Mead, C. A. (1966). Solid-State Electron. 9, 1023. Sects. 1.2, 1.3, 3.3, 3.5.2.

Mead, C. A., & McGill, T. C. (1976). Phys. Letter. 58A, 249. Figs. 2.18, 5.1.

Maeda, K., & Kitahara, E. (1998). Appl. Surf. Sci. 130, 925. Sect. 2.6.1, Fig. 2.19.

Michaelson, H. B. (1977). J. Appl. Phys. 48, 4729. S ect. 3.3.

Miedema, A. R., Boer, F. R. de, & Châtel, P. F. de (1973). J. Phys. F 3, 1558. Sect. 3.3.

Miedema, A. R., Chätel de, P. F., & Boer de, F. R. (1980). Physica. 100B, 1. Sect. 3.3.

Mietze, C., Landmann, M., Rauls, E., Machhadani,H., Sakr, S., Tchernycheva, M., Julien, F. H., Schmidt, W. G., Lischka, K., & As, D. J. (2011). Phys. Rev. B 83, 195301. Figs. 5.22, 5.23.

Misra, V., Heuss, G. P., & Zhong, H. (2001). Appl. Phys. Lett. 78, 4166. Fig. 5.30.

Missous, M., & Rhoderick, E. H. (1986). Electron. Lett. 22, 477. Sect. 2.3.

Miura, Y., Fujieda, S., & Hirose, K. (1994). Phys. Rev. B 50, 4893. Figs. 3.11, 5.11.

Miyake, H., Kimoto, T., & Suda, J. (2009). J. Mat. Sci. Forum 615–617, 979. Fig. 5.20.

Miyazaki, S. (2001). J. Vac. Sci. Technol. B 19, 2212. Figs. 5.12, 5.31, 5.32.

Miyazaki, S., Narasaki, M., Suyama, A., Yamaoka, M., & Murakami, H. (2003). Appl. Surf. Sci. 216, 252. Fig. 5.33.

Mönch, W. (1970). phys. stat. sol. 40, 257. Sect. 3.5.1.

Mönch, W. (1986). In *Festkörperprobleme* (Adv. Solid State Physics) Vol. 26, edited by P. Grosse (Vieweg, Braunschweig), p. 67. Sect. 1.2. Chapt. 7.

Mönch, W. (1987). Phys. Rev. Lett. 58, 1260. Chapt 7.

Mönch, W. (1993). *Semiconductor Surfaces and Interfaces*, 1st Ed. (Springer, Berlin). Sect. 1.2, Apendix.

Mönch, W. (1994a). In *Control of Semiconductor Interfaces,* edited by I. Ohdomari, M. Oshima, and A. Hiraki (Elsevier, Amsterdam, 1994), p. 169. Sect. 2.6.3, 5.1.1.

Mönch, W. (1994b). Europhys. Lett. 27, 479. Sect. 2.5.2.

Mönch, W. (1996). J. Appl. Phys. 80, 5076. Chapt. 6. Sects. 1.2, 4.1, 4.2, 5.1.1, 5.1.2, 5.2, 5.2.4, 5.2.5, 5.2.8. Fig. 5.8

Mönch, W, (1999). J. Vac. Sci. Technol. B 17, 1867. Sect. 2.3.

Mönch, W. (2001). *Semiconductor Surfaces and Interfaces*, 3rd Ed. (Springer, Berlin). Sect. 2.3, 3.5.2, 5.1.1, Fig. 7.4.

Mönch, W. (2004). *Electronic Properties of Semiconductor Interfaces* (Springer, Berlin). Sects. 2.1, 2.3, 2.4, 2.5.2, 3.5.2, 5.1.1.

Mönch, W. (2006). *Band Structure Lineup at I-III-VI2 Schottky Contacts and Hetero-structures,* in *Wide-Gap Chalcopyrites*, edited by S. Siebentritt and U. Rau (Springer, Berlin, 2006), p. 9. Sect. 4.1. Tab. 4.1.

Mönch, W. (2008). Appl. Phys. Lett. 93, 172118. Fig. 3.12.

Mönch, W. (2014). Mat. Sci. Semicond. Processing 28, 2. Sect. 3.5.2.,

Mönch, W. (2017). *Electronic Properties of Semiconductor Interfaces*, in *Springer Hand-book of Electronic and Photonic Materials*, 2nd Edition, edited by S. Kasap and P. Capper (Springer, Berlin), p. 175.

Mohammad, S. N., Fan, Z., Botchkarev, A. E., Kim, W., Aktas, O., Salvador, A., & Morkoç, H. (1996). Electron. Lett. 32, 598. Fig. 5.19.

Mohamed, M., Irmscher, K., Janowitz, C., Galazka, Z., Manzke, R., & Fornari, R. (2012). Appl. Phys. Lett. 101, 132106. Figs. 5.25, 7.4.

Molina, A., & Mohney, S. E. (2022). Mater. Sci. Semicond. Processing 148, 106799. Fig. 5.19.

Morgan, B. A., Talin, A. A., Bi, W. G., Kavanagh, K. L., Williams, R. S., Tu, C. W., Yasuda, T., Yasui, T., & Segawa, Y. (1996a). Mater. Chem. Phys. 46, 224. Sect. 2.4, Fig. 2.13.

Morgan, B. A., Ring, K. M., Kavanagh, K. L., Talin, A. A., Williams, R. S., Yasuda, T., Yasui, T., & Segawa, Y. (1996b). J. Appl. Phys. 79, 1532. Figs. 3.11, 5.11.

Mott, N. F. (1938). Proc. Camb. Philos. Soc. 34, 568. Sects. 1.2, 3.1.

Mourad, D. (2012). Phys. Rev. B 86, 195308. Sect. 4.1.

Mourad, D. (2013). J. Appl. Phys. 113, 123705. Sect. 4.1.

Myburg, G., Barnard, W. O., Meyer, W. E., Louw, C. W., van den Berg, N. G., Hayes, M., Auret, F. D., & Goodman, S. A. (1993). Appl. Surf. Sci. 70–71, 511. Fig. 5.5.

Myburg, G., Auret, F. D., Meyer, W. E., Louw, C. W., & Staden van, M. J. (1998) Thin Solid Films. 325, 181. Fig. 5.5.

Myhra, S. (1978). phys. stat. sol. (a) 49, 285. Figs. 6.1, 6.2.

Nagesh, V., Farmer, J. W., Davis, R. F., & Kong, H. S. (1990). Radiation Effects Defects Solids. 112, 77. Fig. 6.3.

Nathan, M., Shoshani, Z., Ashkinazi, G., Meyler, B., & Zolotarevski, O. (1996). Solid-State Electron. 39, 1457. Fig. 5.17.

Nazarzadehmoafi, M., Machulik, S., Neske, F., Scherer, V., Janowitz, C., Galazka, Z., Mulazzi, M., & Manzke, R. (2014). Appl. Phys. Lett. 105, 162104. Fig. 7.4.

Nguyen, N. V., Kirillov, O. A., Wang, W. J. W., Suehle, J. S., Ye, P. D., Xuan, Y., Goel, N., Choi, K. -W., Tsai, W., & Sayan, S. (2008). Appl. Phys. Lett. 93, 082105. Figs. 5.18, 5.29.

Nienhaus, H., Krix, D., & Glass, S. (2007). J. Vac. Sci. Technol. A 25, 950. Fig. 2.18.

Nishimura, T., Kita, K., & Toriumi, A. (2007). Appl. Phys. Lett. 91, 123123. Sect. 2.6.1 Fig. 5.13.

Novikov, V. A. (1994). Russian Phys. J. 37, 1143. Figs. 6.1, 6.2.

Nuhoğlu, Ç., Ayyíldíz, E., Sağlam, M., & Türüt, A. (1998). Appl. Surf. Sci. 135, 350 and private communication. Fig. 5.17.

Nuhoğlu, C., Aydogan, S., & Türüt. A. (2003). J. Vac. Sci. Technol. A 18, 642. Fig. 2.18.

Nur, O., Sardela, M. R, jr., Willander, M., & Turan, R. (1995). Semicond. Sci. Technol. 10, 551. Figs. 3.11, 5.11.

Oder, T. N., & Naredla, S. B. (2022). AIP Advances. 12, 025117. Fig. 5.15.

Özdemir, A. F., Türüt, A., & Kökçe, A. (2006). Semicond. Sci. Technol. 21, 298. Fig. 5.17.

Ohashi, T., Holmström, P., Kikuchi, A., & Kishino, K. (2006). Appl. Phys. Lett. 89, 041907. Figs. 5.20, 5.24.

Ohdomari, I., Tu, K. N., d'Heurle, F. M, Kuan, T. S., & Petersson, S. (1978). Appl. Phys. Lett. 33, 1028. Sect. 2.4. Figs. 3.11, 5.11.

Ohmi, T., Morita, M., & Hattori, T. (1988) in *The Physics and Chemistry of SiO2 and the Si-SiO2 Interface*, edited by C. R. Helms and B. E. Deal (Plenum, New York, NY,), p. 413. Fig. 5.12.

Ohta, A., Yamaoka, M., & Miyazaki, S. (2004). Microelec. Eng. 72, 154. Figs. 5.12, 5.31, 5.32.

Ohta, A., Nakagawa, H., Murakami, H., Higashi, S., & Miyazaki, S. (2006). e-J. Surf. Sci. Nanotech. 4, 174. Fig. 5.13.

Oishi, T., Koga, Y., Harada, K., & Kasu, M. (2015). Appl. Phys. Express 8, 031101. Fig. 5.25.

Olbrich, A., Vaneca, J., Kreupl, F., & Hoffmann, H. (1997). Appl. Phys. Lett. 70, 2559. Sect. 2.4, Fig. 2.11.

Olbrich, A., Vaneca, J., Kreupl, F., & Hoffmann, H. (1998). J. Appl. Phys. 83, 358. Sect. 2.4, Fig. 2.11.

Oshima, M., Toyoda, S., Okumura, T., Okabayashi, J., Kumigashira, H., Ono, K., Niwa, M., Ohta, A., Nakagawa, H., Murakami, H., Higashi, S., & Miyazaki, S. (2006). e-J. Surf. Sci. Nanotech. 4, 174. Fig. 5.13.

Oshima, T., Kato, Y., Kobayashi, E., & Takahashi. K. (2018). Jpn. J. Appl. Phys. 57, 080308. Figs. 5.26, 5.27, 5.29.

Osvald, J., Kuzmik, J., Konstantinidis, G., Lobotka, P., & Georgakilas, A. (2005). Microelec-tron. Engineer. 81, 181. Fig. 5.19.

Owen, M. H. S., Guo, C., Chen, S. -H., Wan C. -T., Cheng, C. -C., Wu, C. -H., Ko, C. -H., Wann, C. H., Ivana, Zhang, Z., Pan, Ji S., & Yeo, Y. -C. (2013). Appl. Phys. Lett. 103, 031604. Fig. 5.13.

Owen, M. H. S., Guo, C., Chen, S. -H., Wan C. -T., Cheng, C. -C., Wu, C. -H., Ko, C. -H., Wann, C. H., Ivana, Zhang, Z., Pan, Ji, S., & Yeo, Y. -C. (2013). Appl. Phys. Lett. 103, 031604. Fig. 5.13.

Ozbek, M. A., & Baliga, B. J. (2011). Solid State Electron. 62, 1. Fig. 5.19.

Pa, S., Singh, S. D., Dixit, V. K., Sharma, T. K., Kumar, R., Sinha, A. K., Sathe, V., Phase, D. M., Mukherjee, C., & Ingale, A. (2015). J. Alloys Compounds 646, 393. Fig. 5.13.

Paggel, J. J., Neuhold, G., Haak, H., & Horn, K. (1998). Surf. Sci. 414, 221. Figs. 3.11, 5.11.

Palm, H., Arbes, M., & Schulz, M. (1993). Phys. Rev. Lett. 71, 2224. Figs. 3.11, 5.11.

Paradzah, A. T., Auret, F. D., Legodi, M. J., Omotoso, E., & Diale, M. (2015). Nucl. Instrum. Meth. Phys. Res. B 358, 112. Fig. 6.3.

Park, C. Y., An, K. S., Kim, J. S., Park, R. J., Chung, J. W., Kinoshita, T., Kakizaki, A., & Ishii, T. (1995). Phys. Rev. B 52, 8198. Sect. 3.5.2.

Park, K. -B., Ding, Y., Pelza J. P., Neudeck, P. G., & Trunek, A. J. (2006). Appl. Phys. Lett. 89, 042103. Fig. 5.2.

Parui, S., Atxabal, A., Ribeiro, M., Bedoya-Pinto, A., Sun, X., Llopis, R., Casanova, F., & Hueso, L. E. (2015). Appl. Phys. Lett. 107, 183502. Fig. 3.11.

Pauling, L. N. (1939/1960). *The Nature of the Chemical Bond* (CorneII Univ., Ithaca, NY). Sects. 3.3, 3.5.1, 3.5.2.

Peacock, P. W., & Robertson, J. (2002). J. Appl. Phys. 92, 4712. Sect. 5.2.10. Fig. 4.1.

Penn, D. R. (1962). Phys. Rev. 128, 2093. Sects. 1.2, 4.1, 4.2.

Perego, M., Seguini, G., & Fanciulli, M. (2006). J. Appl. Phys. 100, 093718. Fig. 5.13.

Perego, M., Scarel, G., Fanciulli, M., Fedushkin, I. L., & Skatova, A. A. (2007). Appl. Phys. Lett. 90, 162115. Fig. 5.13.

Perego, M., Seguini, G., Scarel, G., Fanciulli, M., & Wallrapp, F. (2008). J. Appl. Phys. 103, 043509. Fig. 5.29.

Petkov, A., Mishra, A., Cattelan, M., Field, D., Pomeroy, J., & Kuball, M. (2023). Scientific Reports 13, 3437. Figs. 5.26, 5.27, 5.31, 5.32.

Petti, D., Cantoni, M., Rinaldi, C., & Bertacco, R. (2011). J. Phys. -Conf. Ser. 292, 012010. Fig. 5.13.

Pickard, G. W. (1906). US Patent 836,531. Sect. 1.1.

Ping, A. T., Schmitz, A. C., Khan, M. A., & Adesida, I. (1996). Electron. Lett. 32, 68. Fig. 5.19.

Poganski, S. (1952). Z. Elektrochemie 56, 193. Sect. 1.3.

Poganski, S. (1953) Z. Physik 134, 469. Sect. 1.3.

Polyakov, A. Y., Smirnov, N. B., Govorkov, A. V., Kozhukhova, E. A., Luo, B., Kim, J., Mehandru, R., Ren, F., Lee, K. P., Pearton, S. J., Osinsky, A. V., & Norris, P. E. (2002). Appl. Phys. Lett. 80, 3352. Figs. 5.16, 5.20.

Polyakov, A. Y., Smirnov, N. B., Govorkov, A. V., Markov, A. V., Pearton, S. J., Kolin, N. G., Merkurisov, D.I., Boiko, V. M., Lee, C. -Ro, & Lee, In-H. (2007). J. Vac. Sci. Technol. B 25, 436. Figs. 6.1, 6.2.

Pong, W., & Paudyal, D. (1981). Phys. Rev. B 23, 3085. Sect. 4.3., Tab. 5.2, Figs. 1.5, 5.35.

Pons, D., & Bourgoini, J. C. (1985). J. Phys. C - Solid State Phys. 18, 3839. Fig. 6.4.

Powell, R. J., & Beairsto, R. C. (1973). Solid-State Electron. 16, 265. Fig. 5.30.

Preobrajenski, A. B., Schömann, S., Gebhardt, R. K., & Chassé, T. (2000). J. Vac. Sci. Techol. B 18, 1973. Fig. 5.13.

Presser, E. (1925) Funkbastler 44, 558. Sect. 1.1.

Presser, E. (1926) DRP 501228. Sect. 1.1.

Qiao, D., Yu, L. S., Lau, S. S., Redwing, J. M., Lin, J. Y., & Jiang, H. X. (2000). J. Appl. Phys. 87, 801. Fig. 5.19.

Rahmatallahpur, S., Yegane, M. (2011). Physica B 406, 1351 Fig. 3 11

Ran, J. X., Liu, B. Y., Ji, X. L., Fariza, A., Liu, Z. T., Wang, J. X., Gao, P., & Wie, T. B. (2020). J. Phys. D - Appl. Phys. 53, 404003. Fig. 5.21.

Renault, O., Barrett, N. T., Samour, D., & Quiais-Marthon, S. (2004). Surf. Sci. 566–568, 526. Fig. 5.12.

Reddy, V. R., Ramesh, C. K., & Choi, C. -J. (2006). Phys. Status Solidi A 203, 622. Fig. 5.19.

Reddy, M. S. R., Kumar, A. A., & Reddy, V. R. (2011). Thin Solid Films 519, 3844. Fig. 5.19.

Reddy, N. N. K, & Reddy, V. R. (2012). Bull. Mater. Sci. 35, 53. Fig. 5.19.

Reddy, P., Bryan, I., Bryan, Z., Tweedie, J., Kirste, R., Collazo, R., & Sitar, Z. (2014). J. Appl. Phys. 116, 194503. Fig. 5.21.

Reddy, P. R. S., Janardhanam, V., Shim, K. -H., Reddy, V. R., Lee, P., Se. N., Choi, C. -J. (2020) Vacuum 171, 109012. Fig. 5.25.

Renault, O., Barrett, N. T., Samour, D., & Quiais-Marthon, S. (2004). Surf. Sci. 566–568, 526. Figs. 5.31, 5.32.

Rhoderick, E. H., & Williams, R. H. (1988). *Metal-Semiconductor Contacts,* 2nd Ed. (Clarendon, Oxford). Sect. 2.1.

Richardson, O. W. (1914). Phil. Mag. 28, 633. Sect. 2.1.

Rickert, K. A., Ellis, A. B., Himpsel, F. J., Lu, H., Schaff, W., Redwing, J. M., Dwikusuma, F., & Kuech, T. F. (2003). Appl. Phys. Lett. 82, 3254. Sect. 5.2.8. Fig.5.3.

Rizzi, A., Lantier, R., Monti, F., Lüth, H., Sala, F. D., Di Carlo, A., & Lugli, P. (1999). J. Vac. Sci. Technol. B 17, 1674). Figs. 5.16, 5.20, 5.22, 5.23.

Robertson, J. (2013). J. Vac. Sci. Technol. A 31, 050821. Sect. 5.2.12.

Robertson, J., & Chen, C. W. (1999). Appl. Phys. Lett. 74, 1168. Sect. 4.1.

Robertson, J., & Falabretti, B. (2006). J. Appl. Phys. 100, 014111. Sect. 4.1, 5.2.10. 5.2.11, 5.2.13. Figs. 4.8, 5.2.9.

Roccaforte, F., Francesco La Via, F., Raineri, V., Pierobon, R., & Zanoni, E. (2003). J. Appl. Phys. 93, 9137. Fig. 5.15.

Roccaforte, F. Giannazzoa, F., Alberti, A., Spera, M., Cannas, M., Corad, I., Pecz, B., Iucolanoe, F., & Greco, G. (2019). Mater. Sci. Semicond. Processing 94, 164. Fig. 5.19.

Roccaforte, F., Greco, G., Fiorenza, P., Di Franco, S., Giannazzo F., La Via, F., Zielinski, M., Mank, H., Jokubavicius, V., & Yakimova, R. (2022). Appl. Surf. Sci. 606, 154896. Fig. 5.15.

Rohlfing, M., Krüger, P., & Pollmann, J. (1993). Phys. Rev. B 48, 17791, and private communications (1995). Sect. 4.1, Fig. 4.2.

Rohlfing, M., Krüger, P., & Pollmann, J. (1995). Phys. Rev. Lett. 75, 3489, and and private communications (1995). Sect. 4.1., Fig. 4.2.

Roul, B., Bhat, T. N., Kumar, M., Rajpalke, M. K., Kalghatgi, A. T., &y Krupanidhi, S. B. (2012). Phys. Status Solidi A 209, 1575. Fig. 5.19.

Ruckh, M., Schmid, D. H., & Schock, H. W. (1994). *J. Appl. Phys.* 76, 5945. Fig. 5.13.

Rumaiz, A. K., Woicik, J. C., Carini, G. A., Siddons, D. P., Cockayne, E., Huey, E., Lysaght, P. S., Fischer, D. A., & Genova, V. (2010). Appl. Phys. Lett. 97, 242108. Fig. 5.13.

Saglam, M., Cimilli, F. E., & Türüt A. (2004). Physica B 348, 397. Sect. 2.6.1, Fig. 2.19, 3.11.

Sarpatwari, K., Mohney, S. E., & Awadelkarim, O. O. (2011). J. Appl. Phys. 109, 014510. Fig. 5.19.

Sasaki, S., Kawahara, K., Feng, G., Alfieri, G., & Kimoto, T. (2011). J. Appl. Phys. 109, 013705. Fig. 6.3.

Satoh, M., & Matsuo, H. (2006). Mater. Sci. Forum 527–529, 923. Fig. 5.2.

Sawada, T., Ito, Y., Imai, K., Suzuki, K., Tomozawa, H., & Sakai, S. (2000). Appl. Surf. Sci. 159–160, 449. Fig. 5.19.

Sayan, E., Garfunkel, E., & Suzer, S. (2002). Appl. Phys. Lett. 80, 2135. Figs. 5.12, 5.31, 5.32.

Schafranek, R., & Klein, A. (2006). Solid State Ionics 177, 1659. Fig. 7.4.

Schleife, A., Fuchs, F., Rödl, C., Furthmüller, J., & Bechstedt F. (2009). Appl. Phys. Lett. 94, 012104. Sects. 1.2, 4.1, 4.2, 5.2.8. Fig. 4.5.

Schlüter, M. (1978). Phys. Rev. B 17, 5044. Sect. 1.2.

Schmitsdorf, R. (1993). Diploma Thesis, Universität Duisburg (unpublished). Sect. 2.3, Fig. 2.3.

Schmitsdorf, R., Kampen, T. U., & Mönch, W. (1995). Surf. Sci. 324, 249. Sects. 2.3, 2.5, Figs. 2.4, 2.6, 2.7, 3.11, 5.11.

Schmitz, A. C., Ping, A. T., Khan, M. A., Chen, Q., Yang, J. W., & Adesida, I. (1998). J. Electron. Matter. 27, 255. Fig. 5.19.

Schottky, W. (1914). Physik. Zeitschr. 15, 872. Sect. 2.1.

Schottky, W. (1938). Naturwissenschaften 26, 843. Sect. 1.1.

Schottky, W. (1939). Z. Physik 113, 367. Sect. 1.1.

Schottky, W. (1940). Physik. Zeitschr. 41, 570. Sects. 1.1, 3.1.

Schottky, W. (1942). Z. Physik 118, 539. Sects. 1.1, 2.1, 2.2.

Schottky, W., & Deutschmann, W. (1929). Physik. Zeitschr. 30, 839. Sects. 1.1, 2.1.

Schottky, W., & Spenke, E. (1939). Wiss. Veröff. Siemens-Werke 18, 225. Sect. 3.1.

Schottky, W., Störmer, R., & Waibel, F. (1931). Z. f. Hochfrequenztechnik 37, 162. Sect. 1.1.

Schuster, A. (1874). Phil. Mag. (4) 48, 251. Sect. 1.1.

Schweickert, H. (1939). Verhandl. D. Phys. Gesell. 3, 99. Sects. 1.2, 3.1.

Sefaoğlu, A., Duman, S., Doğan, S., Gürbulak, B., Tüzemen, S., & Türüt, A. (2008). Micro-electron. Engin. 85, 631. Fig. 5.15.

Sehgal, B. K., Bhattacharya, B., Vinayak, S., & Gulati, R. (1998). Thin Solid Films 330, 146. Fig. 5.17.

Seo, K. -I., McIntyre, P. C., Sun, S., Lee, D. -I., Pianetta, P., & Saraswat, K. C. (2005). *Appl. Phys. Lett.* 87, 042902. Fig. 5.13.

Shammas, J., Yang, Yu, Wang, X., Koeck, F. A. M., McCartney, M. R., Smith, D. J., & Nemanich, R. J. (2017). Appl. Phys. Lett. 111, 171604. Fig. 5.10.

Sheoran, H., Tak, B. R., Manikanthababu, N., & Singh, R. (2020). ECS J. Solid State Sci. Technol. 9, 055004. Fig. 5.25.

Shi, K., Liu, X. L., Li, D. B., Wang, J., Song, H. P., Xu, X. P., Wei, H. Y., Jiao, C. M., Yang, S. Y., Song, H., Zhu, Q. S., & Wang, Z. G. (2011a). Appl. Surf. Sci. 257, 8110. Figs. 5.20, 5.10.

Shi, K., Li, D. B., Song, H. P., Guo, Y., Wang, J., Xu, X. Q., Liu, J. M., Yang, A. L., Wei, H. Y., Zhang, B., Yang, S. Y., Liu, X. L., Zhu, Q. S., & Wang, Z. G. (2011b). Nanoscale Res. Lett. 6, 50. Fig. 5.10, 5.24.

Shigiltchoff, O., Bai, S., Devaty, R. P., Choyke, W. J., Kimoto, T., Hobgood, D., Neudeck, P. G., & Porter, L. M. (2003). Mater. Sci. Forum 433–436, 705. Fig. 5.15.

Shih, C. F., Chen, N. C., Chang, P. H., & Liu, K. S. (2005). Jpn. J. Appl. Phys. Part 1 44, 7892. Fig. 5.20, 5.24.

Shive J. N. (1940). See Bardeen, J. (1947). Sect. 3.1.

Siemens, W. (1875). Monatsber. Berliner Akad. Wissensch. p. 259. Sect. 1.1.

Singh, S. D., Ajimsha, R. S., Sahu, V., Kumar, R., Misra, P., Phase, D. M., Oak, S. M., Kukreja, L. M., Ganguli, T., & Deb, S. K. (2012). Appl. Phys. Lett. 101, 212109. Fig. 5.13.

Sitnitsky, I., Garramone, J. J., Abel, J., Xu, P., Barber, S. D., Ackerman, M. L., Schoelz, J. K., Thibadoa, P. M., & LaBellab, V. P. (2012). J. Vac. Sci. Technol. B 30, 04E110. Fig. 5.5.

Skromme, B. J., Luckowski, E., Moore, K., Bhatnagar, M., Weitzel, C. E. Gehoski, T., & Ganser, D. (2000). J. Electron. Mater. 29, 376. Fig. 5.15.

Song, Y. P., Meirhaeghe van, R. L., Laflère, W. H., & Cardon, E. (1986) Solid-State Electronics 29, 633. Sect. 2.4.

Sorifi, S., Kaushik, S., Sheoran, H., & Singh, R. (2022). J. Phys. D: Appl. Phys. 55, 365105. Fig. 5.12.

Sousa Pires, J. de, Ali, P., Crowder, B., D'Heurle, F., Petersson, S., Stolt, L., & Tove, P. A. (1979). Appl. Phys. Lett. 35, 202. Figs. 3.11, 5.11.

Spicer, W. E., Gregory, P. E., Chye, P. W., Babaola, J. A., & Sukegawa, T. (1975). Appl. Phys. Lett. 27, 617. Fig. 5.17.

Spicer, W. E., Chye, P. W., Skeath, P. R., Su, C. Y., & Lindau, I. (1979). J. Vac. Sci. Technol. 16, 1422. Sect. 1.2.

Spiga, S., Wiemer, C., Tallarida, G., Scarel, G., Ferrari, S., Seguini, G., & Fanciulli, M. (2005). Appl. Phys. Lett. 87, 112904, Fig. 5.13.

Splith, D., Müller, S., Schmidt, F., von Wenckstern, H., van Rensburg, J. H., Meyer, W. E., & Grundmann, M. (2014). Phys. Status Solidi A 211, 40. Fig. 5.25.

Stievenard, D., Boddaert, X., Bourgoin, J. C., & von Bardeleben, H. J. (1990). Phys. Rev. B 41, 5271. Fig. 6.4.

Stockman, L., & van Kempen, H. (1998). Surf. Sci. 408, 232. Fig. 5.17.

Storasta, L., Bergman, J. P., Janzén, E., & Lu, H. J. (2004). J. Appl. Phys. 96, 4909. Fig. 6.3.

Suezaki, T., Kawahito, K., Hatayama, T., Uraoka, Y., & Fuyuki, T. (2001). Jpn. J. Appl. Phys. 40, L43. Fig. 5.15.

Sun, H., Castanedo, C. G. T., Liu, K., Li, K. -H., Guo, W., Lin, R., Liu, X., Li, J., & Li, X. (2017a). Appl. Phys. Lett. 111, 162105. Figs. 5.22, 5.23, 5.26, 5.27.

Sun, L., Lu, H. -L., Chen, H. -Y., Wang, T., Ji, X. -M., Liu, W. -J., Zhao, D., Devi, A., Ding, S. -J., & Zhang, D. W. (2017b). Nanoscale Res. Lett. 12, 102. Figs. 5.22, 5.23.

Sun, S. -M., Liu, W. -J., Wang, Y. -P., Huan, Ya-W., Ma, Q., Zhu, B., Wu, Su-D., Yu, W. -J., Horng, R. -H., Xia,C. -T., Sun, Q,-Q., Ding, S,-J., & Zhang, D. W. (2018a). Appl. Phys. Lett. 113, 031603. Figs. 5.26, 5.27.

Sun, S. -M., Liu, W. -J., Xiao, Yi-F., Huan, Ya-W., Liu, H., Ding, S. -J., & Zhang D. W. (2018b). Nanoscale Res. Lett, 13, 412. Figs. 5.26, 5.27.

Supardan, S. N., Das, P., Major, J. D., Hannah, A., Zaidi, Z. H., Mahapatra, R., Lee, K. B., Valizadeh, R., Houston, P. A., Hall, S., Dhanak, V. R., & Mitrovic, I. Z. (2020). J. Phys. D: Appl. Phys. 53, 075303. Fig. 5.20.

Suri, R., Kirkpatrick, C, J., Lichtenwalner, D. J., & Misra, V. (2010). Appl. Phys. Lett. 96, 042903. Figs. 5.16, 5.29.

Suzuki, R., Nakagomi, S., Kokubun, Y., Arai, N., & Ohira, S. (2009). Appl. Phys. Lett. 94, 222102. Fig. 5.25.

Swaminathan, S., Sun, Y., Pianetta, P., & McIntyre, P. C. (2011). J. Appl. Phys. 110, 094105. Fig. 5.13.

Szydlo, N., & Poirer, R. (1971). J. Appl. Phys. 42, 4880. Fig. 5.28.

Takayanagi, K., Tanishiro, Y., Takahashi, M., & Takahashi, S. (1985). Surf Sci. 164, 367. Sect. 2.5.1, Fig. 2.15.

Tang, M. -Y., Xu, R., Gao, Y. -C., & Wang, L. -J. (2011). J. Shanghai Univ. 15, 218. Fig. 5.10.

Tanner, C. M., Choi, J., & Chang, J. P. (2007). J. Appl. Phys. 101, 034108. Figs. 5.16, 5.29.

Tejedor, C., & Flores, F. (1978). J. Phys. C 11, L19. Sects. 1.3. 3.8.2, 4.1.

Tejedor, C., Flores, F., & Louis, E. (1977). J. Phys. C - Solid State Phys. 10, 2163. Sects. 1.2, 3.6, 3.7, 4.1. Tab. 4.2.

Tersoff, J. (1984a). Phys. Rev. Lett. 52, 465. Sects. 1.2, 4,1, 5.1.2, 5.2, 5.3.2. Fig. 5.8. Tab. 4.2.

Tersoff, J. (1984b). Phys. Rev. B 30, 4874. Sect. 4.1.

Tersoff, J. (1985). Phys. Rev. B 32, 6968. Sect. 4.1.

Tersoff, J. (1986). J. Vac. Sci. Technol. B 4, 1066. Sects. 1.2, 4.1. Tab. 4.2.

Therrrien, F., Zakutayev, A., & Stevanovic, V. (2021). Phys. Rev. Applied 6, 064064. Sect. 5.2.9. Chapt. 7.

Thomas, R. E., Gibson, J. W., & Haas, G. A. (1980). Appl. Surf. Sci. 5, 398. Fig. 7.4.

Tong, X., Ohuchi, S., Tanikawa, T., Harasawa, A., Okuda, T., Aoyagi, Y., Kinishita, T., & Hasegawa, S. (2002). Appl. Surf. Sci. 190, 121. Fig. 3.10.

Topping, J. (1927). Proc. Roy. Soc. A 114, 67. Sect. 3.5.1.

Torvik, J. T., Leksono, M., Pankove, J.I., van Zeghbroeck, B., Ng, H. M., & Moustakas, T. D. (1998). Appl. Phys. Lett. 72, 1371. Figs. 5.20, 5.16.

Toumi, S., Ouennoughi, Z., & Weiss, R. (2021). Appl. Phys. A 127, 661. Fig. 5.15.

Tung, R. T. (1992). Phys. Rev. B 45, 13509. Sect. 2.4, Fig. 2.10.

Turan, R., Aslan, B., Nur, O., Yousif, M. Y. A., & Willander, M. (2001). Appl. Phys. A 72, 587. Figs. 3.11, 5.11.

Ueda, K., Kawamoto, K., Soumiya, T., & Asano, H. (2013). Diamond Related Mater. 38, 41. Figs. 2.18, 5.1.

Ueno, K., Shibahara, K., Atsushi Kobayashi, A., & Fujioka, H. (2022). Appl. Phys. Lett. 118, 022102. Fig. 5.3.

Usuda, K., & Hirashita, N. (2003). Appl. Phys. Lett. 83, 2172. Fig. 5.12.

Varley, J. B., Weber, J. R., Janotti, A., Van de Walle, C. G. (2024). J. Appl. Phys. 135, 075703. Sects. 1.2, 4.1, Fig. 4.11, Table 4.1.

Veal, T. D., King, P. D. C., Hatfield, S. A., Bailey, L. R., McConville, C. F., Martel, B., Moreno, J. C., Frayssinet, E., Semond,F., & Zúñiga-Pérez, J. (2008). Appl. Phys. Lett. 93, 202108. Figs. 5.22, 5.23.

Vereecken, P. M., & Searson, P. C. (1999). Appl. Phys. Lett. 75, 3135. Fig. 5.5.

Vrijmoeth, J., Veen, J. F. van der, Heslinga, D. R., & Klapwijk, T. M. (1990). Phys. Rev. B 42, 9598. Sect. 1.2.

Waag, A., Wu, Y. S., BickneU-Tassius, R. N., Gonser-Buntrock, C., & Landwehr, G. (1990). J. Appl. Phys. 68, 212. Fig. 5.18.

Wagner, L. F., Young, R. W., & Sugerman, A. (1983). IEEE Electron Device Lett. 4, 320. Sect. 2.4.

Wahab, Q., Karlsteen, M., Nur, O., Hultman, L., Willander, M., & Sundgren, J. -E. (1996). J. Electron. Mater. 25, 1495. Fig. 5.12.

Walukiewicz, W. (1988). Phys. Rev. B 37, 4760. Sect. 1.2.

Walukiewicz, W. (2001). Physica B 302–303, 123. Sect. 1.2. Chapt. 7.

Wan, X. -J. Wang, X. -L., Xiao, H. -L., Wang, C. -M., Feng, C., Deng, Q. -W., Qu, S. -Qi, Zhang, J. -W., Hou, X., Cai, S. -J., & Feng, Z. -H. (2013). Chin. Phys. Lett. 30, 057101. Fig. 5.20.

Wang, X. J., & He, L. (1998). J. Electron. Mater. 27, 1272. Fig. 5.19.

Wang, W.I., & Stern, F. (1985). J. Vac. Sci. Technol. B 3, 1280. Fig. 5.18.

Wang, L., Nathan, M.I., Lim, T. -H., Khan, M. A., & Chen, Q. (1996). Appl. Phys. Lett. 68, 1267. Fig. 5.19.

Wang, S. J., Huan, A. C. H., Foo, Y. L., Chai, J. W., Pan J. S., Li, Q., Dong, Y. F., Feng, Y. P., & Ong C. K. (2004). Appl. Phys. Lett. 85, 4418. Figs. 5.12, 5.13.

Wang, K., Lian, C., Su, N. & Jena, D. (2007). Appl. Phys. Lett. 91, 232117. Figs. 5.20, 5.24.

Wang, X. J., Tari, S., Sporken, R., & Sivananthan, S. (2011). Appl. Surf. Sci. 257, 3346. Fig. 5.18.

Wang, X., Xiang, J., Wang, W., Zhang, J., Han, K., Yang, H., Ma, X., Zhao, C., Chen, D., & Ye, T. (2013). Appl. Phys. Lett. 102, 031605. Figs. 5.31, 5.32.

Wang, X, Xiang, J., Wang, W., Zhao, C., & Zhang, J. (2016). Surf. Sci. 651, 94. Fig. 5.13.

Wang, Qi-L., Fu, S. Y., He, Si-H., Zhang, H. -Bo, Cheng, S. -H., Li, L. -An, & Li, H. -D. (2022). Chinese Phys. B 31, 088104. Fig. 5.10.

Watanabe, H., Kirino, T., Kagei, Y., Harries, J., Yoshigoe, A., Teraoka, Y., Mitani, S., Nakano, Y., Nakamura, T., Hosoi, T., & Shimura, T. (2011). Mater. Sci. Forum 679–680, 386. Figs. 5.31, 5.32.

Wei, Q. Y., Li, T., Huang, J. Y., Ponce, F. A., Tschumak, E., Zado, A., & As, D. J. (2012a). Appl. Phys. Lett. 100, 142108. Fig. 5.20.

Wei, W., Qin, Z., Fan, S., Li, Z., Shi, K., Zhu, Q., & Zhang, G. (2012b). Nanoscale Res. Lett. 7, 562. Figs. 5.26, 5.27.

Weidner, M., Frank, T., Pensl, G., Kawasuso, A., Itoh, H., & Krause-Rehberg, R. (2001) Physica B. 308–310, 633. Fig. 6.3.

Welker, H. J. (1943). See Handel, K. Chr. (1999). *Anfänge der Halbleiterforschung und - entwicklung*; Dissertation, RWTH Aachen. Sect. 3.1.

Welker, H. J. (1979). Ann. Rev. Mater. Sci. 9, 1. Sect. 1.1.

Werner, J. H., & Güttler, H. H. (1991). J. Appl. Phys. 69, 1522 (1991). Sect. 2.4.

Werner, P., Jäger, W., & Schüppen, A. (1993). J. Appl. Phys. 74, 3846. Sect. 2.5 Fig. 2.17.

Weyers, S. J., Janzen, O., & Mönch, W. (1999). J. Phys.: Condens. Matter 11, 8489. Sect. 3.5.1. Figs. 3.11, 5.11.

Wheeler, V. D., Shahin, D.I., Tadjer, M. J., & Eddy Jr., C. R. (2017). ECS J. Solid State Sci. Technol. 6, Q3052. Figs. 5.26, 5.27.

Williams, R. (1965). Phys. Rev. 140, A569. Fig. 5.12.

Wilson, A. H. (1931). Proc. Roy. Soc. A 133, 458 and 134, 277. Sect. 1.1.

Wittmer, M. (1991). Phys. Rev. B 43, 4385. Fig. 2.20.

Wohlleben, K., & Beck, W. (1966). Z. Naturforschg. 21a, 1057. Figs. 6.1, 6.2.

Wu, C.I., & Kahn, A. (2000). Appl. Surf. Sci. 162–163, 250. Fig. 7.4.

Wu, C. -L., Lee, H. -M., Kuo, C. -T., & Gwo, S. (2007). Appl. Phys. Lett. 91, 042112. Figs. 5.20, 5.24.

Wu, C. -L., Lee, H. -M., Kuo, C. -T., Chenand, C. -H., & Gwo, S. (2008). Appl. Phys. Lett. 92, 162106. Fig. 5.20, 5.24.

Xiang, J., Wang, G., Li, T., Cui, H., Wang, X., Xu, G., L i, J., Wang, W., & Zhao, C. (2013). ECS-Trans. 58, 153. Fig. 5.29.

Xiong, G., Shao, R., Droubay, T. C., Joly, A. G., Beck, K. M., Chambers, S. A., & Hess, W. P. (2007). Adv. Funct Mater. 17, 2133. Fig. 7.4.

Xu, F., Vos, M., Sullivan, J. P., Atanasoska, L. J., Anderson, S. G., & Weaver, J. H. (1998). Phys. Rev. B 38, 7832. Fig. 5.13.

Xu, K., Sio, H., Kirillov, O. A., Dong, L., Xu, M., Ye, P. D., Gundlach, D., & Nguye, N. V. (2013). J. Appl. Phys. 113, 024504. Fig. 5.28.

Xu, Y., Chen, X., Cheng, L., Ren, F. -F., Zhou, J., Bai, S., Lu, H., Gu, S., Zhang, R., Zheng, Y., & Ye, J. (2019). J. Chinese Phys. B 28, 038503. Fig. 5.25.

Yadav, M. K., Mondal, A., Das, S., Sharma, S. K., & Bag, A. (2020). J. Alloys Comp. 819, 153052. Figs. 5.12, 5.26, 5.27.

Yamamoto, T., Taoka, N., Ohta A., Truyen, N. X., Yamada, H., Takahashi, T., Ikeda, M., Makihara, K., Nakatsuka, O., Shimizu, M., & Miyazak, S. (2018). Jpn. J. Appl. Phys. 57, 06KA05. Figs. 5.26, 5.27.

Yamaoka, M., Narasaki, M., Murakami, H., & Miyazaki, S. (2002). 2nd Internatl. Semicond. Technol. Conf. (Jpn. Electrochem. Soc., Tokyo), p. 229. Fig. 5.12.

Yang, F., Ban, D., Fang, R., Xu, S., Xu, P., & Yuan, S. (1996). J. Electron Spectrosc. Relat. Phenom. 80, 193. Fig. 5.13.

Yang, M., Wu, R. Q., Chen, Q., Deng, W. S., Feng, Y. P., Chai, J. W., Pan, J. S., & Wang, S. J. (2009a). Appl. Phys. Lett. 94, 142903. Fig. 5.13.

Yang, A. L., Song, H. P., Liu, X. L., Wei, H. Y., Guo, Y., Zheng, G. L., Jiao, C. M., Yang, S. Y., Zhu, Q. S., & Wang, Z. G. (2009b). Appl. Phys. Lett. 94, 052101. Figs. 5.22, 5.23.

Yang, A. L., Song, H. P., Wie, H. Y., Liu, X. L., Wang, J., Lv, X. Q., Jin, P., Yang, S. Y., Zhu, Q. S., & Wang, Z. G. (2009c). Appl. Phys. Lett. 94, 163301. Fig. 5.24.

Yang, J., Eller, B. S., Zhu, C., England, C., & Nemanich, R. J. (2012). J. Appl. Phys. 112, 053710. Fig. 5.29.

Yang, J., Ahn, S., Ren, F., Khanna, R., Bevlin, K., Geerpuram, D., Pearton, S. J., & Kuramata, A. (2017). Appl. Phys. Lett. 110, 142101. Fig. 5.25.

Yang, J., Sparks, Z., Ren, F., Pearton, S. J., & Yang, M. T. (2018a). J. Vac. Sci. Technol. B 36, 061201. Fig. 5.25.

Yang, J., Ren, F., Pearton, S. J., & Kuramata, A. (2018b). IEEE Trans. Electron Devices 65, 2790. Fig. 5.25.

Yang, T. -H., Fu, H., Chen, H., Huang, X., Montes, J., Baranowski, I., Fu, K., & Yang, Y. Z. (2019a). J. Semicond. 40, 012801. Fig. 5.25.

Yang, H., Qian, Y., Zhang, C., Wuu, D. -S., Talwar, D. N., Lin, H. -H., Lee, J. -Fu, Wan, L., He, K., & Feng, Z. C. (2019b). Appl. Surf. Sci. 479, 1246. Figs. 5.26, 5.27, 5.29.

Yang, Xu, Pristovsek, M., Nitta, S., Liu, Y., Honda, Y., Koide, Y., Kawarada, H., & Amano, H. (2020). ACS Appl. Mater. Interfaces 12, 46466. Fig. 5.10.

Yang, R.-Y., Cao, Xi-Y., Ma, H.-P., Wen, X.-H., Zhao, X.-F., Yang, L., Shen, Yi (2024). Optical Mater. 150, 115097. Figs. 5.14, 5.26, 5,27.

Yao, Y., Gangireddy, R., Kim, J., Das, K. K., Davis, R. F., & Porter, L. M. (2017). J. Vac. Sci. Technol. B 35, 03D113. Fig. 5.25.

You, J. B., Zhang, X. W., Song, H. P., Ying, J., Guo, Y., Yang, A. L., Yin, Z. G. Chen, N. F., & Zhu Q. S. (2009). J. Appl. Phys. 106, 043709. Figs. 5.31, 5.32.

You, J. B., Zhang, X. W., Zhang, S. G.,Tan, H. R., Ying, J., Yin, Z. G., Zhu,Q. S., & Chu, P. K. (2010). J. Appl. Phys. 107, 083701. Fig. 5.12.

Yu, X., Raisanen, A., Haugstad, G., Ceccone, G., Troullier, N., & Franciosi, A. (1990). Phys. Rev. B 42, 1872. Fig. 5.18.

Yu, X., Raisanen, A., Haugstad, G., Troullier, N., Biasiol, G., & Franciosi, A. (1993). Phys. Rev. B 48, 4545. Fig. 5.13.

Yu, L. S., Qiao, D. J., Xing, Q. J., Lau, S. S., Boutros, K. S., & Redwing, J. M. (1998). Appl. Phys. Lett. 73, 238. Fig. 5.19.

Yu, H. Y., Li, M. F., Cho, B. J., Yeo, C. C., Joo, M. S., Kwong,D. -L., Pan, J. S., Ang, C. H., Zheng, J. Z., & Ramanathan, S. (2002). Appl. Phys. Lett. 81, 376. Figs. 5.12, 5.29.

Yuan, L., Zhang, H., Jia, R., Guo, L., Zhang, Y., & Zhang, Y. (2018). Appl. Surf. Sci. 433, 530. Figs. 5.26, 5.27 5.29.

Zaima, S., Kojima, J., Hayashi, M., Ikeda, H., Iwano, H., & Yasuda, Y. (1995). Jpn. J. Appl. Phys. 34, 741. Fig. 2.18.

Zekentes, K., Kayiambaki, M., & Constantinidis, G. (1995). Appl. Phys. Lett. 66, 3015. Fig. 6.3.

Zhang, S. B., Cohen, M. L., & Louie, S. G. (1986). Phys. Rev. B 34, 768. Sect. 3.5.2.

Zhang, J., Storasta, L., Bergman, J. P., Son, N. T., & Janzén, E. (2003). J. Appl. Phys. 93, 4708. Fig. 6.3.

Zhang, R., Zhang, P., Kang, T., Fan, H., Liu, X., Yang, S., Wei, H., Zhu, Q., & Wang, Z. (2007). Appl. Phys. Lett. 91, 162104. Fig. 5.24.

Zhang, B. L., Cai, F. F., Sun, S., Fan, H. B., Zhang, P. F., Wei, H. Y., Liu, X. L., Yang, C. Y., Zhu, Q. S., & Wang, Z. G. (2008a). Appl. Phys. Lett. 93, 072110. Fig. 5.16.

Zhang, R., Guo, Y., Song, H., Liu, X., Yang, S., Wei, H., Zhu, Q., & Wang, Z. (2008b). Appl. Phys. Lett. 93, 122111. Fig. 5.18, 5.24.

Zhang, W. F., Nishimula, T., Nagashio, K., Kita,K., & Toriumi, A. (2013). Appl. Phys. Lett. 102, 102006. Fig. 5.13.

Zhang, H., Jia, R., Lei, Y., Tang, X., Zhang, Y., & Zhang, Y. (2018). J. Phys. D - Appl. Phys. 51, 075104. Figs. 5.26, 5.27.

Zhang, D., Lin, W., Liu, S., Zhu, Y., Lin, R., Zheng, W., & Huang, F. (2019). ACS Appl. Mater. Interfaces 11, 48071. Fig. 5.25.

Zhang, J., Han, S., Cui, M., Xu, X., Li, W., Xu, H., Jin, C., Gu, M., Chen, L., & Zhang, K. H. L. (2020a). ACS Appl. Electron. Mater. 2, 456. Figs. 5.26, 5.27.

Zhang, Ya-Z., Li, Yi F., Wang, Z. Z., Guo, R., Xu, S. R. Liu, C. -Y., Zhao, S. L., Zhang, J. C., & Ha, Y. (2020b). Sci. China-Phys. Mech. Astron. 63, 117311. Figs. 5.26, 5.27.

Zhao, G., Li, H., Wang, L., Meng, Y., Li, F., Wei, H., Yang, S., & Wang, Z. (2018). Appl. Phys. A 124, 130. Figs. 5.22, 5.23.

Zhi, Y., Liu, Z., Wang, X., Li, S., Wang, X., Chu, X., Li, P., Guo, D., Wu, Z., & Tang, W. (2020). J. Vac. Sci. Technol. A 38, 023202. Figs. 5.26, 5.27.

Zhou, P., Spencer, M. G., Harris, G. L., & Fekade, K. (1987). Appl. Phys. Lett. 50, 1384. Fig. 6.3.

Zhou, Yi, Wang, D., Ahyi, C., Tin, C. -C., Williams, J., Par, M., Williams, N. M., Hanser, A., & Preble, E. A. (2007). J. Appl. Phys. 101, 024506. Fig. 5.19.

Zhou, Q., Wu, H., Li, H., Tang, X., Qin, Z., Dong, D., Lin, Y., Lu, XC., Qiu, R., Zheng, R., Wang, J., & Li, B. (2019). J. Electron Devices Soc. 7, 662. Fig. 5.21.

Zhu, S., Detavernier, C., Meirhaeghe, R. L. van, Cardon, F., Blondeel, A., Clauws, P., Ru, G. -P., & Li, B. -Z. (2001). Semicond. Sci. Technol. 16, 83. Figs. 3.11, 5.11.

9783031590634